Ecosystem Functioning

In the face of decreasing biodiversity and ongoing global changes, maintaining ecosystem functioning is seen both as a means to preserve biological diversity and for safeguarding human well-being by securing the services ecosystems provide. The concept today is prominent in many fields of ecology and conservation biology, such as biodiversity research, ecosystem management or restoration ecology. Although the idea of ecosystem functioning is important, the concept itself remains rather vague and elusive. This book provides a novel analysis and integrated synthesis of different approaches to conceptualising and assessing ecosystem functioning. It links the natural sciences with methodologies from philosophy and the social sciences, and introduces a new methodology for a clearer and more efficient application of ecosystem functioning concepts in practice. Special emphasis is laid on the social dimensions of the concept and on the ways in which these influence research practice. Several case studies relate theoretical analyses to practical application.

KURT JAX is a Senior Scientist at the Helmholtz Centre for Environmental Research – UFZ – in Leipzig, Germany, and Professor of Ecology at the Department of Ecology and Ecosystem Management of the Technische Universität München. His research focuses on the conceptual foundations of ecology and conservation biology, with a special emphasis on the application of ecological concepts as tools for conservation biology and on the adaptation of methods from the humanities (especially philosophy) for interdisciplinary research in the environmental sciences.

ECOLOGY, BIODIVERSITY AND CONSERVATION

The world's biological diversity faces unprecedented threats. The urgent challenge facing the concerned biologist is to understand ecological processes well enough to maintain their functioning in the face of the pressures resulting from human population growth. Those concerned with the conservation of biodiversity and with restoration also need to be acquainted with the political, social, historical, economic, and legal frameworks within which ecological and conservation practice must be developed. The new *Ecology, Biodiversity and Conservation* series will present balanced, comprehensive, up-to-date, and critical reviews of selected topics within the sciences of ecology and conservation biology, both botanical and zoological, and both 'pure' and 'applied'. It is aimed at advanced final-year undergraduates, graduate students, researchers, and university teachers, as well as ecologists and conservationists in industry, government, and the voluntary sectors. The series encompasses a wide range of approaches and scales (spatial, temporal, and taxonomic), including quantitative, theoretical, population, community, ecosystem, landscape, historical, experimental, behavioural, and evolutionary studies. The emphasis is on science related to the real world of plants and animals rather than on purely theoretical abstractions and mathematical models. Books in this series will, wherever possible, consider issues from a broad perspective. Some books will challenge existing paradigms and present new ecological concepts, empirical or theoretical models, and testable hypotheses. Other books will explore new approaches and present syntheses on topics of ecological importance.

Ecology and Control of Introduced Plants
Judith H. Myers and Dawn Bazely

Invertebrate Conservation and Agricultural Ecosystems
T. R. New

Risks and Decisions for Conservation and Environmental Management
Mark Burgman

Ecosystem Functioning

KURT JAX

Helmholtz Centre for Environmental Research (UFZ),
Leipzig, Germany

CAMBRIDGE
UNIVERSITY PRESS

CAMBRIDGE
UNIVERSITY PRESS

Shaftesbury Road, Cambridge CB2 8EA, United Kingdom

One Liberty Plaza, 20th Floor, New York, NY 10006, USA

477 Williamstown Road, Port Melbourne, VIC 3207, Australia

314–321, 3rd Floor, Plot 3, Splendor Forum, Jasola District Centre, New Delhi – 110025, India

103 Penang Road, #05–06/07, Visioncrest Commercial, Singapore 238467

Cambridge University Press is part of Cambridge University Press & Assessment, a department of the University of Cambridge.

We share the University's mission to contribute to society through the pursuit of education, learning and research at the highest international levels of excellence.

www.cambridge.org
Information on this title: www.cambridge.org/9780521705233

First published 2010

A catalogue record for this publication is available from the British Library

Library of Congress Cataloging-in-Publication data
Jax, Kurt, 1958–
Ecosystem functioning / Kurt Jax.
    p.   cm. – (Ecology, biodiversity, and conservation)
Includes bibliographical references and index.
ISBN 978-0-521-87953-8 – ISBN 978-0-521-70523-3 (pbk.)
1. Biotic communities.   2. Ecosystem management.   I. Title.
QH541.J39   2010
577 – dc22   2010029487

ISBN   978-0-521-87953-8   Hardback
ISBN   978-0-521-70523-3   Paperback

"Solche Missgriffe, setzte er abbrechend hinzu, sind unvermeidlich, seitdem wir von dem Baum der Erkenntnis gegessen haben. Doch das Paradies ist verriegelt und der Cherub hinter uns; wir müssen die Reise um die Welt machen, und sehen, ob es vielleicht von hinten irgendwo wieder offen ist."
[. . .]
"Mithin, sagte ich ein wenig zerstreut, müssten wir wieder von dem Baum der Erkenntnis essen, um in den stand der Unschuld zurückzufallen? Allerdings, antwortete er, das ist das letzte Kapitel von der Geschichte der Welt."

Heinrich von Kleist: *Über das Marionettentheater* (1810)

"Such blunders," he added, interrupting himself, "are unavoidable, since we have eaten of the tree of knowledge. But Paradise is locked and bolted and the Cherub is behind us. We must journey around the world, to see if a back door has perhaps been left open."
[. . .]
"Would that mean," I said somewhat absentminded, "that we would have to eat of the tree of knowledge once more to fall back into the state of innocence?" "Indeed" he answered, "and that is the final chapter in the history of the world."

Heinrich von Kleist: *On the Marionette Theater* (1810)

Contents

Acknowledgements

Many of the ideas this book deals with have occupied my thinking for a long time. So I was very excited when Michael Usher asked me to write a book on the topic of ecosystem functioning, as it allowed me to draw together, synthesise, and extend these thoughts. It nevertheless took much longer than anticipated because I had to venture into some fields, such as restoration ecology, the European Water Framework Directive, or the debate about constructivism, that were rather new to me.

My ideas about ecological units, ecosystem functioning, and ecology in general have developed over many years. They benefited from discussions with many colleagues and have also been furthered by various experiences in the field, some of which form the basis of case studies in this book. So a great many people and institutions have contributed over the years in one way or another to what is written in this book – so many that I can name only some of them here.

The SIC model, described in Section 4.4, was first developed together with my colleagues, Steward Pickett and Clive Jones, Millbrook, New York, who I thank for many intense and enjoyable discussions, and for their great openness in pursuing together a notion with an – at first – rather uncertain outcome. Some parts of Section 4.4 follow closely the text of our common publication (Jax *et al.*, 1998). They are here reprinted with permission of Oikos, as are some selected paragraphs in Sections 4.1 and 4.2, which are taken from Jax (2005).

A fellowship at the DFG-postgraduate programme, 'Ethics in the Sciences and Humanities', at the University of Tübingen not only allowed me to delve deeper into the field of environmental ethics, but also provided an opportunity to apply my theoretical considerations on ecological units (ecosystems, communities, populations) to a specific case study, namely ecosystem management in Yellowstone National Park. My special thanks here go to the staff of the Yellowstone Research Library, then still at Mammoth Hot Springs, in particular Lee Whittlesey. My research in Yellowstone National Park was also generously supported by a Robert

Whittaker Fellowship of the Ecological Society of America. The results of this research form the basis of the Yellowstone case study in Section 4.3. The text of that section is a slightly modified and updated translation of a paper of mine that was previously published in German as Jax (2001), and which is reprinted here with permission of the publisher, namely the Verlag für Wissenschaft und Bildung (VWB), Berlin.

My development of the theoretical ideas in this book have also greatly benefited from many discussions with my students, who, in several seminars about the concepts of function in general and ecosystem functioning in particular, inspired and supported me with their critical questions.

Thanks go also to my colleagues at the Technische Universität München, Freising, and the UFZ-Helmholtz Centre for Environmental Research, Leipzig, for countless productive discussions and conversations.

I am also very grateful to Ricardo Rozzi (Puerto Williams, Chile, and Denton, Texas), who opened up to me the opportunity of doing research at the 'end of the world', namely the Chilean island of Navarino, which is the focus of my first case study (Chapter 2). This beautiful place has provided a magnificent and inspiring experience for me. My travels to Chile and my work there have been supported by travel grants from the German Academic Exchange Service (DAAD) and the Fondo de las Americas; some of my research and that of my students has been supported through the German–Chilean research project BIOKONCHIL, funded by the German Ministry of Education and Research (BMBF), FKZ 01LM0208.

Drafts of some chapters of the book were read by Harald Auge, Uta Berghöfer, Fridolin Brand, Christoph Görg, Carsten Neßhöver, Irene Ring, Ursula Schmedtje, Elke Schüttler, Ludwig Trepl, Michael Usher, and Angela Weil. Their comments were invaluable to me and greatly helped to improve the book.

My thanks go also to all those who provided illustrations for this book; specific credits are given in the respective figure captions.

Finally, I want to express my thanks to Michael Usher and to the team at Cambridge University Press, who were always highly responsive to my questions and helpful regarding my requests – and patient when the manuscript took much longer to complete than originally planned.

1 · Introduction

Concern about the functioning of the world's ecosystems has become commonplace, in the scientific literature as well as in everyday parlance. Climate change, loss of biological diversity, chemical pollution, land use changes, and the spread of exotic species are all discussed in connection with the perceived or anticipated degradation or destruction of ecosystems, or at least with an impairment of their functioning. While attention has been focused in the past mostly on the fate of specific processes relevant to human life (such as clean water or the maintenance of food production) or specific valued species, the emphasis has shifted increasingly towards a broader perspective, namely that of the whole ecosystem. Since about the early 1990s, ecosystems and their functioning have become major targets of conservation and management, accompanied by biodiversity as the other major broad-scale conservation focus. Today, both conservation aims are embodied in national and international management strategies, such as the variety of ecosystem management approaches (e.g. Yaffee *et al.*, 1996; Boyce and Haney, 1997) or the Convention on Biological Diversity (including also an 'Ecosystem Approach' as a cross-cutting issue), and the various regional and national strategies that are still newly developed. These trends have also triggered a large amount of scientific research related to these fields, which vice versa reinforced the political processes. The concept (or at least the term) 'ecosystem functioning' (also 'ecosystem function') has thus become a major topic of ecological research during the last decades, especially in connection with biodiversity research.

The notion of focusing not on single ('sectoral') aspects of the environment and/or ecological systems but rather on the performance of whole ecosystems appears to be a useful approach. In the face of limited resources for research and management measures and under the conditions of accelerating global changes, it seems a wise decision to concentrate our efforts on understanding and maintaining whole systems – and hopefully at the same time saving the maximum components of these systems (especially the species which constitute them) (Walker, 1992, 1995), as well as the services they provide for human wellbeing (Daily, 1997; Millennium Ecosystem Assessment, 2003, 2005). If there is something like ecosystems, and not just a happenstance or loose collection of relations between organisms and their environment, we should be able to speak of these systems as either functioning or not functioning, or about different degrees of functioning – as is the common parlance in ecology and conservation biology. We consider ecosystems as collapsing (e.g. when forests die off, lakes turn to hypertrophic states, or coral reefs bleach), we speak of degraded ecosystems, whose functioning is impaired (e.g. when leaching or soil erosion take place, when exotic species begin to dominate an ecosystem, or when some populations of fish become extinct), and we talk about intact, complete, and functioning systems (e.g. talking about the Greater Yellowstone Ecosystem as 'one of the largest, relatively intact temperate zone ecosystems left on earth' (Glick *et al.*, 1991, p. 9)).

Measuring and managing ecosystem functioning should thus provide a powerful and far-reaching tool for the management of nature (and its services). At the same time it should greatly contribute to our understanding of the ecological theatre (to use a term of the late G. E. Hutchinson), both through the basic research initiated by these practical concerns (see Srivastava and Vellend, 2005) and by the practical experiences gained from monitoring management results.

While the idea of ecosystem functioning is intuitively highly appealing to most people – including myself – implementing the conservation of ecosystems and their functioning is far from trivial. Some people would even argue that the concept of ecosystem functioning is not a useful concept at all, because it might rest on an unscientific view of goals inherent in nature (or ecosystems specifically). In any case, questions must begin with the understanding of the concept itself: do we all mean the same thing when we talk about ecosystem functioning or about a functioning ecosystem? Are we addressing the same object when we look at 'the' ecosystem? How can we measure ecosystem functioning? What (and who) decides if an ecosystem is functioning? When is an ecosystem destroyed? How do we evaluate different states of ecosystems? How do we arrive at reference states for functioning ecosystems? How do other concepts (such as ecosystem integrity or ecosystem health) relate to the notion of ecosystem functioning? To what degree is ecosystem functioning a descriptive concept and to what degree a normative concept?

These are the questions this book will follow. It is thus a book on the concept of ecosystem functioning and the way this concept is or can be put into practice. As we will see, this will carry us far beyond the boundaries of empirical studies of ecology and conservation biology, into questions about the theory of ecosystems and about the interface between ecology, philosophy, and the social sciences. Although great emphasis will be laid on what we know about the functioning of ecosystems (and this functioning's relations to biodiversity), this is not a textbook on ecosystems as such. But neither is it a book on the philosophy of ecosystems and ecosystem functioning. Instead, it attempts to build a bridge between theoretical and practical issues, the latter ranging from field research through management. The book is thus, in the first place, targeted at ecologists and conservation biologists.

The structure of the book

Investigating ecosystem functioning is commonly considered an issue of ecosystem research and thus of the natural sciences. A second look at the topic, however, reveals that it is not only about ecology, but also requires taking into account some ideas and tools that are normally seen as belonging to the humanities. While thinking about ecosystem functioning, questions about epistemology, teleology, and norms pop up. As I will show, such tools are not just supplemental for when all the science is done, they are essential for conceptualising and operationalising ecosystem functioning. Thus, the arguments in this book will oscillate between the natural sciences and the humanities. Being a biologist, I see my own role not in purely analysing the process of doing science nor in developing new philosophical tools. I see myself more as *applying* philosophical tools to a new and specific subject – the issue of ecosystem functioning. By these means I hope to sharpen our ecological concepts and methods (and tools as well), and with it our understanding of the living world and how it may be preserved. I thus feel (surprisingly for myself) in the line of my own family tradition – that is, more a craftsman of philosophy than an artist in this field.

The first chapters of this book are devoted to an analysis of the different uses and various meanings of ecosystem functioning, and to the conceptual and practical issues related to them. The later chapters will then try to synthesise the ideas described. By means of an array of different conceptual tools, the chapters describe the necessary steps (as I see them) towards putting the concept of ecosystem functioning into practice. To aid the understanding of the conceptual intricacies of the notion, I will draw heavily on case studies, which can exemplify the difficulties and requirements of implementing ecosystem functioning.

This is what Chapter 2 starts with. A case study dealing with research conducted on the impact of the introduced Canadian beaver on the ecosystems of Navarino Island (southern Chile) (Section 2.1) will be used to introduce some of the major contexts of the use of ecosystem functioning concepts. Section 2.2 will then further broaden the context, describing in brief some other major fields in which the concept is of importance, such as restoration ecology or ecosystem management. Chapter 3 is devoted to the relationship between biodiversity and ecosystem functioning. This is the area where the issue is currently discussed most intensively, at least with respect to using the *term* 'ecosystem functioning'. As we will see, there are two fundamentally different ways in which the term is used. 'Ecosystem functioning' either refers to selected processes and properties at the ecosystem level, or to the overall performance (or operation) of the whole system. I will investigate the scientific discourse on biodiversity and ecosystem functioning in some more detail and analyse which variables are selected here as measures for ecosystem functioning. One basic question will be to what degree these variables are and can be considered as proxies for the notion of overall ecosystem performance – as is often implied, or at least alluded to. Or, are these variables and the processes or properties to which they relate of interest on their own?

It will become evident that there is some sort of terminological and conceptual confusion when talking about ecosystem functioning. This is only in part a matter of terminological convention. More than that, some philosophical problems hover in the background of the implicit mixing of different meanings of ecosystem functioning. These problems – and how they can be resolved – are dealt with in Chapter 4. I will start with an elaboration of the different meanings of 'function' in the environmental sciences – as the root of the word 'functioning' – and differentiate at least four different meanings of the term (Section 4.1). Although most of the following text will focus on the meaning of ecosystem function(ing) as the 'overall performance of ecosystems' (ecosystem functioning proper, or ecosystem functioning in the narrow sense), all of the other major meanings (function as process, function as services, function as roles) will nevertheless play important roles in understanding and implementing ecosystem functioning. Talking about function and functioning in biology is always subject to critical questions as to whether it implies teleological thinking, i.e. thinking in terms of goals within nature (or ecosystems in our case). At least some modes of teleological descriptions are seen as highly problematic within science. A huge literature exists with respect to the relation between function concepts and teleological thinking, mostly related to the concepts of the organism, the gene, or (human) society. To avoid pitfalls and unnecessary controversies in the application of ecosystem functioning concepts, it is necessary to ponder the implications of these discussions for our topic. This will be done in Section 4.2. As I will show, not every use of seemingly teleological assumptions within concepts of ecosystem functioning is really problematic. Some, however, are, and I will sketch ways to avoid them.

We then have to proceed to the other part of the expression 'ecosystem functioning', namely to the 'ecosystem' (Section 4.3). There is an even greater variety of meanings here than we find when analysing 'function'. This variety has important implications for our ability to agree on a measure of what ecosystem functioning is and – in applying the concept – what constitutes a functioning ecosystem. Far from analysing the many conceptual differences of the ecosystem concept, which has been done in other places (see Jax, 2006), I will first demonstrate the causes and consequences of different ecosystem concepts for implementing ecosystem management. The long and fascinating history of ecosystem management in Yellowstone National Park will serve as a very telling example. In connection with this case study, I will introduce a conceptual model (called the SIC model, explained in Chapter 4) to communicate and clarify the many possible definitions of an ecosystem. This model will be used later (Chapter 7) as one possible tool for operationalising ecosystem functioning.

The Yellowstone example also demonstrates how scientific and non-scientific ideas (such as our ideas of what nature is) intermingle in ecology and conservation biology. With this, it becomes clear that another (sometimes implicit) part of most of the uses of the ecosystem functioning concept is the notion of functioning as *proper* functioning. This means that describing ecosystem functioning not only requires clarification of the concept of the ecosystem implied, but also of a reference state for a functioning ecosystem, in comparison to which it is judged as either functioning (properly) or not functioning.

Having assembled the parts of a definition of ecosystem functioning so far as 'the overall performance of an ecosystem as compared to a reference state or dynamic', we can finally open up our view to include a variety of related concepts, which form together what I call a conceptual cluster. That is, we are not dealing with the *term* 'ecosystem functioning', but with the *concept* as coarsely defined above. Related concepts thus are: ecosystem resilience, ecosystem integrity, ecosystem health; but also ecosystem collapse, ecosystem reliability, etc.

The results of what is discussed in Chapter 4 raise new questions, which refer to the relationship between science and society when defining and assessing ecosystem functioning. This is the topic of Chapter 5. First, if there are many different possibilities for conceptualising ecosystems, are ecosystems mere mental constructs and not 'real'? Are they – to refer to a very heated controversy about the status of science and nature – only socially constructed? Would, then, ecosystem functioning also be merely a social construct and thus be completely relative? I will argue in Section 5.1 for a moderate realism, acknowledging the necessary social dimensions of all concept and theory formation while nevertheless not denying the existence of something 'out there' which resists a completely deliberate and artificial definition of ecosystems and ecosystem functioning, at least when these concepts are applied in practice. As one classical example of discontinuities in the overall performance of ecosystems, I will describe the discussion about alternative stable states in shallow lakes. Nevertheless, many different and (for specific purposes) appropriate ways of defining ecosystem functioning remain, and with them the question of who determines what a functioning ecosystem is. For this reason, part of the chapter will also deal with (and refute) the argument that acknowledging the partial epistemological relativity of defining ecosystems (and thus ecosystem functioning) would imply or foster *moral* relativity towards environmental problems.

In order to better understand the character of the social dimension in the definition of ecosystem functioning, I will, in Section 5.2, describe the various types of value decisions the selection of an (appropriate) concept of ecosystem functioning involves. This leads to the important question of normative dimensions of concepts like ecosystem functioning. When we change – as is frequently done – the order of words from 'ecosystem functioning' to 'functioning ecosystem', the normative dimensions (in terms of a *proper* functioning implied) of the concept come to the fore still stronger. Science, however, by its definition and its modern self-image, should be value-neutral and, as such, should keep clear of normative aspects. This image, however, has been challenged time and again and is not congruent with the practice of science. Particularly in conservation biology, many concepts (and the discipline as such) are highly value-laden. The question is more, which kind of norms and values enter our concepts in which way and how can they be made explicit? Acknowledging and explicating the different normative dimensions of ecosystem functioning is, in fact, less of a problem than it can be a solution to adequately defining and implementing the concept. In another case study (Section 5.3) I will analyse the complex relations between concepts of ecosystem functioning and societal choices in the context of ecosystem management strategies, focusing especially on the Ecosystem Approach of the Convention on Biological Diversity. Section 5.4 will draw some general conclusions on the roles of science and society in assessing ecosystem functioning.

In Chapter 6 I will proceed towards a synthesis. After a summary of the analysis elaborated in the preceding chapters (Section 6.1), some of the more prominent approaches of conceptualising and measuring ecosystem functioning (*sensu stricto*) will be discussed in Section 6.2. These are, especially, ecosystem integrity and ecosystem health, and ecosystem stability and resilience. They will be scrutinised by means of the tools and critical questions developed previously. The major questions applied to all of these approaches are: (a) which definition(s) of an ecosystem they follow; (b) which reference conditions they envisage; and (c) in which way societal choices are included in their definition and implementation. As a case study for the use of an ecosystem integrity concept, I will discuss the European Water Framework Directive and its goal of reaching 'good ecological status' in European surface waters.

While the approaches described in Chapter 6 all have some merits, I do not think they encompass the whole breadth and intuitive understanding of ecosystem functioning. At the beginning of the final chapter I will therefore use the SIC model, introduced in Chapter 4, to set out the possible definitions of ecosystem functioning. The four most common meanings of ecosystem functioning in an empirical setting (as I see them), will be described in more detail (Section 7.1). I will emphasise, in particular, how these meanings can be put into practice, i.e. how ecosystem functioning can be assessed. Determination of whether an ecosystem is functioning, or to what degree it is a functioning ecosystem, will, however, not always coincide. That is, applied to the same chunk of nature (e.g. the Greater Yellowstone Ecosystem or the ecosystem(s) of Navarino Island), an ecosystem will be considered as functioning under one definition but destroyed or 'malfunctioning' under another. As there is no use in searching for 'the' proper definition of ecosystem functioning, this brings us back to the crucial question of how to select an appropriate definition for specific purposes. To this end, Section 7.2 presents some guidelines for conceptualising and assessing ecosystem functioning in conservation practices, as a kind of checklist of necessary choices and procedures. The guidelines are followed by a final case study that illustrates the use of ecosystem functioning concepts in ecological restoration, building on an empirical example from strip mining rehabilitation in Lower Lusatia (Germany). The overall conclusions of the book and an outlook are given in Section 7.4. This final section very briefly draws together the most important results of this book. I then describe what the results mean for research on ecosystem functioning, as well as for the application of the concept, emphasising not least its ethical implications.

There is no unified theory or unified concept of ecosystem functioning. There will be none at the end of this book. But I hope there will be some more clarity about how to formulate and apply unambiguous concepts of ecosystem functioning and how to proceed in building more refined and restricted theories about it.

2 · *Setting the scene*
The context of investigating ecosystem functioning

The issue of ecosystem functioning has been addressed in a variety of different contexts. In this chapter, I will introduce the most important of these contexts. First, however, I will open up a number of questions with respect to the meaning of the term 'ecosystem functioning' and the ways the concept is assessed in practice. To do so, I will start with a case study illustrating the many facets of the current discourse on ecosystem functioning.

2.1 Case study: exotic species and ecosystem functioning on Navarino Island

The Chilean island of Navarino is a remote and beautiful place, and an outstanding site of nature. I first visited the island in 2000, invited by my colleague and friend Ricardo Rozzi, who had started to study the biological and cultural diversity of the island a few years before. Military issues – tensions between Chile and its neighbour Argentina – meant access was largely restricted until the 1990s. Located in the XII Region of Chile, Navarino is one of the numerous islands which, at the southernmost tip of the American continent, form the archipelago of Tierra del Fuego. Being situated south of the Beagle Channel and only around 150 km north of Cape Horn (54° south), it is also a rather cold place. It harbours only a small human population of about 2300 people. Nevertheless, the island and its surroundings have been inhabited for at least 6500 years by the southernmost ethnic group of the world, the Yaghan or Yamana (Gusinde, 1937; Borrero, 1997; Martinic, 2002). Today the descendants of the indigenous Yaghan people, a small group of approximately 70 persons, form the Comunidad Yaghan. The majority of the population is comprised of European/Chilean settlers, the first of which arrived during the late nineteenth and early twentieth centuries, and by soldiers of the Chilean navy and their families. Puerto Williams, the only larger settlement on the island, was founded as a strategic naval

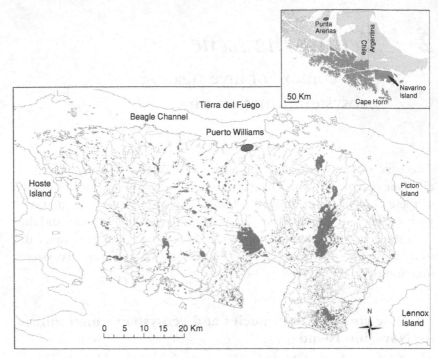

Fig. 2.1. Navarino Island, Chile and its location at the southern tip of South America. The dark-shaded area on the smaller map designates the extension of the Cape Horn Biosphere Reserve. Map courtesy of Elke Schüttler, Puerto Williams.

base in 1953, and is the southernmost permanent settlement in the world. The inhabited area of the island is limited to the coastline, especially the northern coast (Rozzi *et al.*, 2006), where only one (unpaved) road leads along most of the coastline.

Navarino (Fig. 2.1) covers an area of approximately 2500 km². Characteristic of the landscape is the mountain ridge of the 'Dientes de Navarino' ('teeth of Navarino'), whose highest peaks rise to well over 1000 m above sea level. The northern coast is sheltered from southerly storms by the mountain range and large extents of forest have developed there. The larger southern half of Navarino Island is an open expanse of sub-Antarctic tundra, interspersed with occasional patches of trees. Hundreds of small moor ponds and a number of larger, relatively shallow lakes are scattered across this marshy region (Sielfeld and Venegas, 1980).

Due to the insular conditions and the Arctic influence, a subhumid to humid climate has developed in the Tierra del Fuego archipelago

(Lizarralde, 1993). Short, cool summers (mean temperature 9°C) and long, moderately cold winters (mean temperature 2°C) characterise the region. It receives about 500–650 mm of precipitation annually (Moore, 1983; Tuhkanen, 1992).

The forests of the Tierra del Fuego archipelago are the southernmost forest ecosystems in the world. They are part of the Magellanic forests, which are considered one of the most important remaining wilderness areas on the globe (Mittermaier *et al.*, 2003). Due to biogeographical barriers these southernmost forests have an isolated position and are therefore species-poor with regard to most taxa, and are characterised by a high degree of endemism (Moore, 1983; Rebertus and Veblen, 1993). The forests of Navarino Island cover an area of about 800 km^2. They are dominated by three species of the southern beech: *Nothofagus pumilio*, *N. antarctica* (both deciduous) and *N. betuloides* (evergreen), the only other tree-like species being winter's bark (*Drymis winteri*) and pickwood (*Maytenus magellanicus*).

The region is relatively poor in terms of vascular plant species, but possesses a high diversity of landscape types and mosaics of different terrestrial and aquatic ecosystems (Rozzi *et al.*, 2006), comprising forests, moorland, peat bog and other wetlands, shrubland, alpine vegetation, and several coastal forms. This variety of ecosystems goes along with a high diversity of non-vascular plant species which show a great deal of specialisation concerning substrates and habitat types. Especially in terms of non-vascular plants, the area can even be called a hotspot of biodiversity (Rozzi *et al.*, 2008). In 2005, the Cape Horn region, including Navarino Island, was designated as a new UNESCO Biosphere reserve (Rozzi *et al.*, 2006).

However, even though the area is so remote and is described as pristine, there is concern about the functioning of these southern ecosystems. Besides possible impacts of developments such as increased and uncontrolled tourism, a major influence altering the ecosystems of the island is the presence of exotic species. This refers especially to the influence of exotic vertebrates. Among the vertebrates, the mammalian assemblage of Navarino is comprised of 5 native species and 11 exotics, the ecologically most conspicuous exotic species being the North American beaver (*Castor canadensis*), the North American mink (*Neovison vison*), and feral pigs and dogs. The freshwater fish fauna is constituted of only three native species and three exotics. In contrast to mammals, the avifauna is comparatively diverse, with dozens of forest and marine birds (Anderson and Rozzi, 2000; Couve and Vidal-Ojeda, 2000). Only two exotic bird

Box 2.1 *Exotic and invasive species: terminology*

The language of dealing with 'exotic' and 'invasive' species is often value laden and sometimes even highly emotional (Eser, 1999; Woods and Moriatry, 2001; Larson, 2005). There has also been a lot of discussion about how to find appropriate and clear definitions for the terms used in this discourse (e.g. Colautti and MacIsaac, 2004; Pyšek *et al.*, 2004; Heger and Trepl, 2008). I am using the expressions 'exotic' and 'invasive species' here in a more neutral ecologically and less conservation-focused manner. Thus, I define 'exotic species' as species which occur outside of their native range. Following Heger and Trepl (2003, 2008), by 'invasive species' I mean: any species that occurs *and spreads* at a location outside its area of origin; the occurrence of the species must have been prevented in the past by a barrier to dispersal, and not by the conditions in the new habitat (Heger and Trepl, 2003, p. 314, modified). This definition neither presumes human agency in the process of dispersal nor 'negative' impacts with respect to matters of biological conservation. Please note also, that my intention in this text is not to discuss the complex issue of exotic/invasive species as such. I only use it as an example to make my point: what is meant by 'ecosystem functioning' and how it is assessed in practice.

species have been observed, which are also restricted to the village of Puerto Williams (all data on exotic vertebrate species distribution in the area are taken from Anderson *et al.*, 2006a; specifically for fish see also Moorman *et al.*, 2009). No reptiles or amphibians have been observed in the region (Gusinde, 1937; Anderson *et al.*, 2006a).

With respect to the effects of invasive species, a question raised frequently is whether, and if so to what degree, they influence the functioning of (native) ecosystems. It has been stated repeatedly that 'invading species can [...] drastically alter ecosystem functioning' (Parker *et al.*, 1999, p. 8), can have 'dramatic effects on ecosystem functioning' (Carlsson *et al.*, 2004, p. 1575), or are 'a threat to biodiversity and ecosystem functioning' (Jaksic *et al.*, 2002, p. 157). In 1990, Peter Vitousek, drawing on his experiences with a variety of exotic species from Hawaii, stated: 'If an introduced species can in and of itself alter ecosystem-level processes such as primary or secondary productivity, hydrology, nutrient cycling, soil development, or disturbance frequency, then clearly the properties of individual species can control the functioning of whole ecosystems' (Vitousek, 1990, p. 8).

Fig. 2.2. The Canadian beaver (*Castor canadensis*) on Navarino Island. Photo: André Künzelmann, UFZ, February 2006.

More than any other animal, the beaver has attracted the attention of conservationists, biologists, and − due to its clearly visible effects on the landscape − even the broad public and the regional administration. *Castor canadensis* (Fig. 2.2), one of the world's two beaver species, has its original distribution throughout North America, from northern Mexico to the Arctic tundra, except in areas of the arid southwest, peninsular Florida, and the high Arctic (Baker and Hill, 2003; Naiman *et al.*, 1988). The beaver was introduced in Tierra del Fuego in 1946 to promote fur exploitation. Twenty-five mating beaver pairs, trapped in Canada, were released on the Isla Grande of Tierra del Fuego. The absence of most natural predators and the rich abundance of forage and habitats have led to a rapid increase in the beaver population since its introduction (Lizarralde, 1993; Anderson *et al.*, 2009). In 1962, 16 years after their first introduction to Isla Grande, beavers were discovered on the northern coast of Navarino Island for the first time (Sielfeld and Venegas, 1980; Anderson *et al.*, 2009). Until today these animals have occupied almost every suitable habitat on the whole island at a reported density of 1.1 colonies stream km^{-1} (Skewes *et al.*, 2006). Anderson *et al.* (2009) state that the animals occupy 2–15% of the landscape with impoundments and meadows, and affect 40–50% of channel length of the streams on Navarino Island.

What are the effects of the beaver on the island? On the most general level, beavers do on Navarino what they do everywhere: they build dams and cut down shrubs and trees, which they use as building material and food. Two major factors govern where beavers are found: the type of water body and the abundance of food in and immediately around these water bodies. Beavers require a permanent and stable supply of water throughout the year, and they can control water levels on streams, ponds, and lakes by building dams. Beavers are completely herbivorous and their diet varies throughout the year. During summer their preferred diet is made up of herbaceous plants. During winter and early spring, they subsist on the bark and wood of tree species within the surrounding area.

The beaver is the classical example of an 'ecosystem engineer' (Jones *et al.*, 1994, 1997), i.e. an organism creating or modifying habitats ('directly or indirectly control the availability of resources to other organisms by causing physical state changes in biotic or abiotic materials' (Jones *et al.*, 1997, p. 1947)).

Few other animals have the ability to change landscapes as much as the beaver. Naiman *et al.* (1988) assume that changes created by beavers may last for several centuries. As ecosystem engineers, beavers modulate the availability of resources to other species. Beaver activity has consequences for the abiotic and biotic environment. Abiotic consequences are physical, chemical, and geomorphological effects – for example, expansion of wetland areas, elevation of the water table, reduction of current velocity, changes in bed-slope, and increased retention and accumulation of organic and sedimentary material, resulting in changes to the biochemical composition of water, soil, and sediments (Lizarralde, 1993). The main direct biotic effect of the beaver is the felling of the *Nothofagus* trees during foraging and building activities. As you can see in Fig. 2.3, the animals even approach large old trees. By doing so, and by the associated damming activities, it creates new habitats and increases habitat mosaics (Fig. 2.4). This leads to changes in species composition, circulation of matter, and energy flows in the aquatic and terrestrial environment. As a result of that, some species originally living in these places lose their habitat, but on the other hand, the new environmental conditions offer suitable living spaces for other species. Particularly some birds, above all some geese species (on Navarino, for example, *Chloephaga poliocephala* or *Anas flavirostris*), gain due to the new, more open areas along the beaver ponds, which act as preferred breeding places (Skewes *et al.*, 1999).

Beyond its ecological impacts, the beaver is responsible for some social and economic effects on Navarino Island. It can influence forestry and

Fig. 2.3. Nothofagus trees on Navarino Island 'attacked' by beavers. Photo: K. Jax, March 2002.

Fig. 2.4. Recently abandoned beaver pond near Guerrico, Navarino Island. Photo: Courtesy of Romy Werner, January 2005.

Fig. 2.5. Map of Puerto Williams, Navarino Island, with a beaver as a mascot. Photo: Courtesy of Elke Schüttler, April 2005.

livestock management and cause damage to infrastructure, such as inundation or undermining of roads and paths, and can negatively affect the provision of drinking water. On the other hand, the beaver itself can represent an economic object by being hunted for its meat, fur, and secretion products. More than that, on Navarino it is also used as a tourist attraction, due to its activities as well as its meat, which is served in some restaurants of Puerto Williams (and also in the regional capital, Punta Arenas). In Puerto Williams it is, at times, even used as a mascot, happily welcoming visitors to the island on a map of the town near the city hall (Fig. 2.5).

The beaver visibly has major effects on the landscapes of Navarino (and other parts of Tierra del Fuego), especially on its forests and streams. Some years ago, the Chilean Agriculture Agency (Servicio Agrícola y Ganadero (SAG)) started a control programme for the beaver (Servicio Agrícola y Ganadero, 2003) to limit the damage caused by it to timber and infrastructure. In terms of biological conservation, however, the question goes further than just direct damages to infrastructure such as roads and directly used resources such as timber or meadows. Here the question posed is also whether, and to what degree, the beaver affects biodiversity and the functioning of the region's ecosystems. Christopher Anderson, who has studied this question, warns that '[t]he beaver engineering activities in sub-Antarctic riparian forests appear to be facilitating the establishment and spread of exotics, thereby threatening the region's still highly pristine ecosystems and native biodiversity' (Anderson *et al.*, 2006b, p. 473).

So let us take a closer look and see what exactly the effects of the beaver on the ecosystems of Navarino are, and what it may mean that ecosystem functions have been altered and/or that the ecosystem as a whole may be threatened. The question here is thus one about the meaning of the expressions 'ecosystem functions' and 'ecosystem functioning', and how they are assessed in practice, in this specific case.

Anderson and colleagues focused their studies on streams and the adjacent forest. On the one hand, they compared species compositions of riparian forests on four watersheds with that of former beaver ponds (called 'beaver meadows') in the same watersheds (Anderson *et al.*, 2006b). They found that species composition had changed severely. Also, forest re-growth ('regeneration') five years after abandonment by the beaver took place rather slowly or not at all. The tree seedlings or saplings that re-grew on the beaver meadows were predominantly those of the species *Nothofagus antarctica*. This species, which is considered a pioneer of non-forest habitats such as peat bogs and rangeland (Martinéz Pastur *et al.*, 2006), has only minor abundance in forests unaffected by beavers. Anderson *et al.* thus even speak of an 'alternative stable state': '[O]n Navarino Island our results indicated a potentially long-term elimination of trees by suppressing the seedling bank, thus creating an alternative stable state for riparian ecosystems' (Anderson *et al.*, 2006b, p. 472).

Species richness was significantly higher on beaver meadows than in the adjacent riparian forests, both for native and exotic plants. Similar results, also for older beaver meadows, were obtained by Martinéz Pastur *et al.* (2006) (Argentinean part of Tierra del Fuego) and by Werner *et al.* (2009)(Navarino Island). Some of the beaver meadows investigated in these studies were up to 20 years old. How succession will develop in the long run is still open.

In another part of their studies (Anderson and Rosemond, 2007), Anderson and colleagues investigated the influence of the beaver on the streams themselves, comparing active beaver ponds within the watersheds with unaltered stretches above the beaver ponds and stretches downstream of the ponds. Their investigations here focused on in-stream diversity of benthic macroinvertebrates, community patterns of these invertebrates, and ecosystem function. As measures for 'ecosystem function' they quantified macroinvertebrate biomass and secondary production (Anderson and Rosemond, 2007, p. 142).

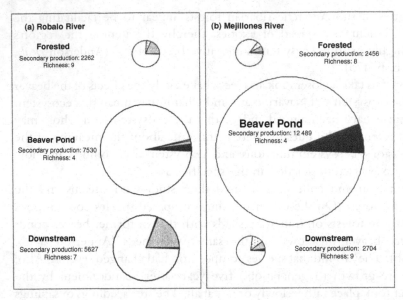

Fig. 2.6. Impact of beavers on the stream ecosystem of two watersheds on Navarino Island. The size of the circles indicates total annual secondary production of benthic macroinvertebrates in 2003. The different shadings show the relative contributions of different functional feeding groups. Graph from Anderson and Rosemond (2007, p. 147); see text and original publication for details. Reprinted with permission from Springer Science + Business Media.

With respect to diversity, they found that it was lowest in pond sites (28 total taxa richness), highest in the 'pristine' upstream areas (33 taxa), and intermediate in the sites downstream of the beaver ponds (31 taxa). This observation, which shows up also in other measures of diversity (e.g. Shannon–Wiener diversity), is attributed to the lower structural diversity of lentic habitats (slow-flowing, here: beaver ponds) as compared to lotic (fast-flowing) ones.

In terms of what they measured as ecosystem function, both benthic biomass and secondary production were highest in the ponds and lower in the downstream and upstream stretches. Depending on the specific stream, either the upstream or the downstream site displayed the lowest values for these variables. In addition, Anderson and Rosemond found a shift in the distribution of the functional types of benthic organisms (see Fig. 2.6), accompanied by an increase of trophic diversity in the beaver ponds (taxa were classified here following the 'classical' functional feeding groups of benthic macroinvertebrates, such as shredder, filterer, or predator, reaching back at least to Cummins (1974; see Section 3.2)).

Box 2.2 *Reference conditions*

A *reference condition* or *reference state* is a defined condition of a system against which another system (or the same system at another time) is being compared. It mostly is a condition that is considered as 'normal' and/or desirable. Reference conditions can be derived in many different ways, either from the state of a similar system (often assumed as being 'natural'), by historical data, by modelling approaches, or by stipulation of a desirable condition from a political point of view. For an ecosystem, reference conditions can be described by many different variables, depending on the purposes of a study. A reference condition might be characterised by specific species compositions, specified physical and chemical parameters (as in the European Water Framework Directive; see below and Section 6.2), specific expressions of some 'typical' species only, or levels of productivity. As ecological systems are rarely equilibrium systems, reference states can never be defined with the precision that can be used in reference to technology. At least some variance around every defined condition must always be taken into account. Moreover, reference conditions do not have to be static, but can also be expressed as *reference dynamics*, i.e. referring to (linear or cyclic) trajectories (e.g. successional ones) an ecosystem is expected to follow. Several examples of how reference conditions are defined are given throughout this book.

Looking at these thoroughly conducted studies, we may now ask what it tells us about the function (or functioning) of the riparian and riverine ecosystems of Navarino Island under the influence of the invasive beaver. Are the ecosystems of Navarino influenced by the beaver still functioning? Do they function differently?

A couple of questions arise with respect to conceptualising and assessing ecosystem functioning: does ecosystem functioning refer to single (ecosystem) processes or to the performance of the whole system? Or, perhaps, are some processes selected as proxies for the performance of the whole system? If yes, why these and not other processes? Would an inference from these processes to the whole system be justified? What is the reference state (or reference dynamic) for the functioning of the whole ecosystem? Which deviations from the reference state would be considered as threatening to the system? Who decides about the reference state and its 'permissible' variation/deviation?

These questions are not unique to the current case study, nor the only ones we must posit here with respect to the meaning and measurement of ecosystem functioning. But these questions are characteristic of the whole conceptual discourse about *ecosystem functioning*. The current case study and the papers published about it are representative of many others dealing with ecosystem functioning. I will thus briefly discuss at least some of these questions for the present case study, but will deal with them in a more general and systematic way again in the chapters that follow.

The first important question here is what we refer to when we talk about ecosystem function and/or functioning (the two expressions are sometimes used synonymously, sometimes they are not): do we refer to some *processes* at the ecosystem *level* or to the 'whole system'? In the papers of Anderson and colleagues (and in many other publications; see Chapter 3) there is evidence that both ecosystem-level processes and the whole system are meant, while some authors make a difference between the two expressions. One might argue that this is an over-subtle distinction and that both expressions refer to the same thing. In fact, as I will show later in more detail, they do not. Speaking about the whole system implies the idea of a unit, with a certain kind of spatial and/or functional coherence and boundaries or even integration, whereas speaking about ecosystem-level processes might merely refer to specific flows of energy and matter. In the latter case, the ecosystem is thus not perceived as a unit, but as a kind of *perspective* (see Jax, 2006; Box 5.1). Vitousek (1990), in his paper on the effects of invasive species in ecosystems, explicitly uses the expressions 'ecosystem processes' and 'ecosystem level process' to characterise those phenomena that relate to matter and energy flows, as distinct from 'the functioning of whole ecosystems' (Vitousek, 1990, p. 8).

For Anderson and Rosemond, *ecosystem function* clearly refers to ecosystem (level) processes and state variables such as secondary production, macroinvertebrate biomass, and nitrogen availability (Anderson and Rosemond, 2007, pp. 142, 149). At the same time, they state that a 'process by which invasive species may impact an *entire ecosystem* is via ecosystem engineering, i.e. creating, destroying or modifying habitat' (Anderson and Rosemond, 2007, p. 142, my emphasis), thus referring to the whole system. This kind of *measuring* of ecosystem-level variables while more or less implicitly *referring* to the whole (entire) system is typical for a large part of the current treatment of ecosystem functioning. This holds also for the assumption that measuring some 'typical' ecosystem processes allows us to infer the fate/functioning/intactness of the whole ecosystem. In the current example this becomes evident from the sentence quoted above about the threat to Navarino's ecosystems. Likewise, in a more recent publication Anderson and colleagues state: '[T]he physical changes caused by beavers, which lead to increased retention of organic matter and increased productivity of benthic organisms, finally result in an overall shift in ecosystem function and processes' (Anderson et al., 2009).

The authors draw these conclusions on the basis of their investigations of only a few selected ecosystem processes as described above.

Such generalisations are quite common in the literature, especially in the discussion of the relationship between biodiversity and ecosystem function(ing), to which Anderson and Rosemond (2007) also refer in the discussion of their results. They conclude:

An underlying assumption when examining the relationships between species diversity and ecosystem functioning is that the loss of species from ecosystems will result in a reduction of ecosystem function (Chapin et al., 2000, Naaem and Wright 2003). While this general concept may likely hold in many ecosystems, particularly in regard to such taxa as foundations species (Ellison et al., 2005), our study showed that reductions in species richness were associated with increased ecosystem function (secondary production). *(Anderson and Rosemond, 2007, p. 149)*

While the authors argue on the basis of the state of the art of this field, their statement reveals a basic theoretical weakness of the discourse (and not specifically of their work!): in practice, 'ecosystem function(ing)' is frequently dealt with as a clear and unified concept – which it is not. Thus, inferences are made time and again from single ecosystem-level processes to other ecosystem processes and/or to the whole system, assuming that the effects that biodiversity has on ecosystem processes can easily be extrapolated to any other ecosystem process (see Section 3.2). Otherwise, it would be necessary to first discuss *under which conditions* results obtained on the correlation of biodiversity and ecosystem processes may be extrapolated, instead of speaking, as happens very commonly, about 'ecosystem function' in general. I will come back to this important issue later.

Let us note that a frequent way of talking about – and especially of measuring – ecosystem functioning is to refer to selected phenomena (processes, states) at the ecosystem level. Mostly, however, there is at least a tacit reference to a whole system, beyond individual ecological processes and states.

As to whether these phenomena are in fact proxies for the functioning of the whole system depends strongly on the implied definition of the ecosystem and on the reference state or reference dynamics. Assuming, for the time being, that we have a clear idea about our system (which, in fact, we often have not; see Section 4.3), there still remains the question of *at which point* deviations from the reference conditions are considered so significant that we would not only talk about some *change* in ecosystem functioning, but of the transition into another stable state of the system, into another system, or even into a complete 'collapse' of the ecosystem and its functioning.

At the end of one of their papers, Anderson and colleagues in fact state that this point of transition has already been reached when they consider the possibility that the beavers' activities in the Cape Horn region is creating a 'novel ecosystem' (Anderson *et al.*, 2006b, p. 469) and/or 'an alternative stable state for sub-Antarctic riparian ecosystems' (Anderson *et al.*, 2006b, p. 473). As a consequence, they demand, from a perspective of biological conservation, not only to manage but also to actively *restore* these riparian ecosystems.

With this brief analysis of the Navarino case study, many of the elements and problems of the notion of ecosystem functioning are already in place: the variety of different meanings of ecosystem functioning; the question of the relationship between ecosystem processes and ecosystem performance as a whole; the multitude of variables and approaches for measuring ecosystem functioning; the need to define reference systems and states; the network of concepts in which investigations about ecosystem functioning are situated; and the connection of the concept to normative questions (what is the *desired* state of the ecosystem?) – especially in a conservation context.

Before I proceed to a step-by-step treatment of these issues in the next chapters, however, I will first take a closer look at the major contexts in which the concept of ecosystem functioning is debated and applied.

2.2 The fields of application for 'ecosystem functioning'

The frequent use of the term 'ecosystem functioning' within the scientific literature is a rather recent phenomenon. Figure 2.7, based on a search on the Web of Science website, displays the trend of the appearance of the term in the literature. It demonstrates that the expression 'ecosystem function(ing)' only became popular in the 1990s. We have to be careful, however, to not only rely on the appearance of the *phrase* 'ecosystem

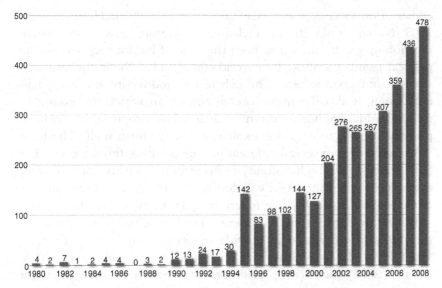

Fig. 2.7. Number of citations per year within the Web of Science containing the term 'ecosystem functioning' or 'ecosystem function' in the search category 'topic'. In addition to the data displayed, there were only occasional entries which referred to these terms before 1980, with a maximum of five in 1978. Date of search: 9 April 2009.

functioning' as an indicator for interest in and research on the issue. The basic ideas behind the phrase have also been dealt with under a couple of different headings. Such headings are (in part): ecosystem stability, ecosystem resilience, ecosystem health, and ecosystem integrity, or simply ecosystem performance – to name only the most prominent ones.

Even though there are many different contexts that in one or the other way refer to ecosystem functioning, there are some – partly overlapping – fields in which the concept is of importance on its own and thus applied. The case study above alludes to the major fields in which 'ecosystem functioning' is an issue of importance, i.e. the major contexts in which the notion of ecosystem functioning is actually used.

The first of these major fields of application is the debate about the relation between biodiversity and ecosystem functioning. In the case study presented above, Anderson and Rosemond (2007) refer to this debate briefly in the discussion of their results, as described. The discourse on if and how changes in biodiversity enhance or decrease ecosystem functioning is one of the major fields in ecology and even more in conservation biology today (see Loreau *et al.*, 2002a; Hooper *et al.*, 2005; Naeem *et al.*, 2009a). Relating biodiversity (loss) to ecosystem functioning greatly helped to keep the issue of biodiversity loss on the political agenda and moved it beyond ethical and aesthetic concerns and into the utilitarian sphere. The debate on biodiversity and ecosystem functioning is also the major overall context in which the notion of ecosystem functioning is currently discussed as a topic on its own, in particular, if one looks at the explicit use of the term itself. The term is, however, used in several different meanings within this discourse. For both reasons – its high visibility in science and society, and the very diverse use of the concept – I will devote an entire chapter to the analysis of the biodiversity–ecosystem functioning debate (Chapter 3).

Largely overlapping, but not identical to this debate is the one about the ability of ecosystems to provide *ecosystem services* (Costanza *et al.*, 1997; Daily, 1997; Millennium Ecosystem Assessment, 2003, 2005; Kremen, 2005), such as the provision of food and fibre, climate and water regulation, or even recreation and spiritual values. Being most strongly expressed in the Millennium Ecosystem Assessment and its follow-up process, ecosystem functioning is perceived here in terms of the usability of ecological systems for the purposes of human wellbeing. Note, however, that ecosystem functioning is sometimes considered a prerequisite for maintaining ecosystem services, while on the other hand, the provision of ecosystem services is also used as a criterion for assessing the functioning of an ecosystem. For more details on the ecosystem services approach and its application, see Sections 4.1 and 7.1.2.

Ecosystem management is a field where we would expect a pronounced treatment of ecosystem functioning. Within the plethora of approaches to ecosystem management and projects to implement it (see e.g. Christensen *et al.*, 1996; Boyce and Haney, 1997), most studies consider not only patterns ('structure') of ecosystems, but also the processes that maintain these structures. For some scientists and managers, these processes are the very essence of ecosystems, sometimes much more important than specific species. This is often expressed as the goal of maintaining the function(ing) of the ecosystems.

The postulate to protect ecosystem functioning has even become part of the guidelines for the management policies of the US National Park Service (NPS): 'The Service will not intervene in natural biological or physical processes, except: [. . .] To restore natural ecosystem functioning that has been disrupted by past or ongoing human activities' (National Park Service, 2006, p. 37).

As we will see in Chapter 4, the idea of ecosystem functioning, in terms of maintaining 'intact' ecosystems, has led to severe controversies in the practice of ecosystem management. I will discuss these controversies and their theoretical background in detail for the example of the wildlife management on Yellowstone's northern range.

'Ecosystem approaches', as methodologies for conservation and resource management moving away from single-species and single-features management towards the management of whole ecosystems, have been established in many fields and by many agencies and organisations. The most popular of these approaches is the Ecosystem Approach of the Convention on Biological Diversity (CBD) (see the case study in Chapter 5), which has also influenced other ecosystem approaches such as that of UNEP, or has been taken up by others, such as UNESCO's Man and the Biosphere Programme (UNESCO, 2000). Other ecosystem approaches are promoted specifically for forest management (see Butler and Koontz, 2005), water management (e.g. in the European Water Framework Directive), and fisheries management (e.g. US Fish and Wildlife Service, 1995). Most of these target the maintenance of ecosystem functioning as one of their major goals. Thus, Principle 5 of the Ecosystem Approach of the CBD states: 'A key feature of the ecosystem approach includes conservation of ecosystem structure and functioning', while Principle 6 postulates: 'Ecosystems must be managed within the limits of their functioning' (see UNEP/CBD, 2000. The so-called Malawi Principles were adopted by the Fifth Conference of the Parties to the CBD in Nairobi, Kenya, in May 2000. For the complete list of the 12 principles, see Section 5.3). Likewise, the European Water Framework Directive (European Community, 2000) defines one of its major terms, 'ecological status', as 'an expression of the quality of the structure and functioning of aquatic ecosystems associated with surface waters' and subsequently in several places refers to the goal of maintaining the functioning of ecosystems (see the case study in Section 6.2 for more details).

Another field in which ecosystem functioning is an explicit issue is *restoration ecology*. In the past, ecological restoration often aimed in the first place at the re-creation of particular species assemblages or landscape features. More recently, the goal of restoration is increasingly seen as restoring ecosystem function(ing). According to the *Primer of the Society for Ecological Restoration International* (Society for Ecological Restoration, 2004), one of the nine attributes of a successfully restored ecosystem is thus: 'The restored ecosystem apparently functions normally for its ecological stage of development, and signs of dysfunction are absent.'

Likewise, in discussing realistic restoration objectives in the face of climate change, Harris et al. (2006, p. 174) conclude: 'Hence, an increasing emphasis will be on proper functioning condition of a site – ecological integrity – and to a lesser extent on nudging a site back to historical conditions based on species. In general, processes, not structures, will prevail.'

Even the (re)introduction of species into an area is not only viewed exclusively in terms of historical species distribution patterns or species protection, but increasingly also as a means to restore ecosystem functioning (Byers et al., 2006; Seddon et al., 2007).

Interestingly, the discourse about ecosystem functioning and restoration ecology has been largely separate from the one on biodiversity and ecosystem functioning (but see Naeem, 2006; Wright et al., 2009), although many possible links exist. I will analyse the use of ecosystem functioning concepts in restoration ecology in a case study in Chapter 7.

A new and emerging field, where ecosystem functioning *should* at least play a decisive role is the research about 'novel ecosystems' (Hobbs et al., 2006, 2009; Seastedt et al., 2008). A novel ecosystem can be defined as a combination of species and abiotic variables that have not existed before in a particular region ('no-analogue systems' (Fox, 2007)). Although such systems have been around in the past (e.g. as a result of agricultural practices), their kind and number will increase substantially in the future as a result of global change (especially climate change and introduced exotic species). As is known from paleoecological studies, species do not shift as whole communities in response to climatic change, but in an individualistic manner (Miles, 1987; Delcourt and Delcourt, 1991). For such systems, questions of what a functioning ecosystem is have to be addressed with even more urgency. While some argue that ecosystem management and ecosystem restoration of 'traditional' ecosystems might use the 'natural' state or at least some historical state (or dynamics, respectively) as a reference condition for assessing their functioning, this is evidently not possible for novel ecosystems. Nevertheless, in analysing and/or managing these ecosystems, we might want – and need – some criterion for their (proper) functioning.

Finally, there is a large field of studies that addresses the issue of if and how ecosystems are functioning (properly) in terms of *the resilience, integrity, or health of ecosystems* (see Rapport, 1995; Gunderson, 2000; Pimentel et al., 2000; Lackey, 2001; Brand and Jax, 2007). Although the term *ecosystem functioning* is not always used here at all, the very notion of a system that is 'exhibiting biological and chemical activities characteristic of its type' (as the Ecological Society of America (1999, p. 4) defined a functioning ecosystem) or the (proper) performance of an ecosystem is clearly the background of all these studies. Clearly defined (either singular or multiple) reference states for a (properly) functioning ecosystem are mostly assumed, as well as thresholds beyond which the ecosystem either malfunctions, moves to another state (alternative stable state; see the case study in Section 5.1), or 'collapses'. There are very specific – sometimes almost in-group – discourses, especially about ecosystem health and ecosystem resilience, with efforts to build an almost all-embracing resilience theory (Gunderson and Holling, 2002; Walker and Salt, 2006). Beyond these discourses, the concepts are also used in the context of the application fields described above. However, while many specific measurement variables for stability, resilience, integrity, and health have been proposed and discussed in the specialised discourses, the terms are used rather vaguely in many other contexts. I will describe the problems and potentials of these approaches to ecosystem functioning in Chapter 6.

In all of the described research fields, both basic and applied research is conducted. The emphasis, however, is clearly on the applied side, following from concerns about the importance of ecosystem functioning for biological conservation and human wellbeing.

2.3 Conclusions from this chapter

As we have seen, 'ecosystem functioning' means different, but related things. A very complex picture and a multitude of questions open up once one dives into the issue of how we talk about ecosystem functioning and how we assess and measure it. There are many fields of application where the concept of ecosystem functioning is relevant, such as research on the role of biodiversity for ecosystems and for humans, ecosystem management, or ecological restoration. For this reason, a deeper analysis of these questions is justified and necessary. As should be clear by now, this book will not just deal with the *term* (or perhaps more correctly, the *phrase*) 'ecosystem functioning' and its different meanings, but with the general *idea* of ecosystem functioning. At least two different facets of this idea became visible in the Navarino case study: ecosystem function(ing) as denoting selected ecosystem processes, and ecosystem function(ing) as the overall performance/operation of an ecosystem, with the former (selected ecosystem processes) often considered to indicate the latter. The meaning of ecosystem functioning as describing the overall performance of an ecosystem is also embodied in other expressions, such as ecosystem integrity, ecosystem health, ecosystem reliability, ecosystem vulnerability, etc. These terms together constitute what I have called a '*conceptual cluster*', that is, an assemblage of concepts with common properties with respect to their epistemology, their meaning and function in ecology, and with respect to the phenomena they describe (Jax, 2006; Jax and Schwarz, 2010; see also Section 4.4).

In the following chapters, I will develop in more detail both the characteristics of the concept 'ecosystem functioning' and the differences of its various expressions. In analysing the debate on the relation between biodiversity and ecosystem functioning, as I will do in the next chapter, I will show how the different meanings and aspects of ecosystem functioning introduced above (as well as some more) are all present, but often in rather woolly and unclear form. I will also provide a first analysis of the variables through which ecosystem functioning is measured and assessed.

3 · *What do we need for a functioning ecosystem?*

The debate on biodiversity and ecosystem functioning

The perhaps most catchy image about relations between biodiversity and ecosystem functioning has been drawn by Paul and Anne Ehrlich in the preface of their book, *Extinction*, which was published in 1981. They describe the different biological species as 'rivets' holding together the wings of an aeroplane (symbolising the earth's ecosystems) in which we ourselves fly. But at the same time, the airline is paying a person (the now infamous 'rivet popper') to pop out the rivets of the wing one by one and sell them for profit. The rivet popper does not seem to be worried very much about what he is doing:

'Don't worry', he assures you. 'I am certain the manufacturer made this plane much stronger than it needs to be, so no harm's done. Besides, I've taken lots of rivets from this wing and it hasn't fallen off yet. [. . .] As a matter of fact, I am going to fly on this flight also, so you can see there's absolutely nothing to be concerned about.' *(Ehrlich and Ehrlich, 1981, p. xi)*

The message the authors want to convey is clear: the ecosystems of our planet are dependent on species and their diversity. While some (or many) species extinctions may go unnoticed, or at least without significantly impairing the functioning of our ecosystems, there may be a threshold in terms of species numbers beyond which the whole system collapses and becomes dysfunctional – the wing will fall off and the plane will crash. A decade after the Ehrlichs' book, Australian ecologist Brian Walker (1992) addressed the question from a slightly different angle, when he introduced the idea of 'ecological redundancy'. While the argument raised by the Ehrlichs in their rivet-popper example proposed that species have to be preserved in order to allow for the sustained existence of our ecosystems as life-support systems, Walker turned the argument around: 'By maintaining the integrity of ecosystem function we minimize the chances of losing the many species we have not yet described and those of whose very existence we are as yet unaware' (Walker, 1992, p. 20).

Here, conserving species (as many as possible) was best achieved by conserving functioning ecosystems. Both approaches, however, in fact converge (Ehrlich and Walker, 1998). In order to maintain ecosystem function(ing), Walker proposes to conserve those species with first priority, which are not redundant, i.e. those species in the ecosystem which fulfil 'functional roles' that are necessary to 'keep the system running' (i.e. functioning) and for which no other functionally similar species are present in the system. These are – in the parlance of the rivet-popper example – those rivets which are indispensable for keeping the wing of the plane intact. Both approaches, in the end, also aim not at the functioning of the systems per se, but at the preservation of species or biological diversity and of the services ecosystems provide for humanity. The Ehrlichs' book was written as an alarm bell: in concern about the ongoing loss of species. In his paper, Walker wanted to find a strategic solution to the seemingly impossible task of conserving all species (or at least as many as possible) while at the same time not having enough knowledge about all species (nor the time and funds necessary to achieve it). While acknowledging that '[t]he worrisome cost of decline in biodiversity, particularly to politicians who may be held accountable, is the threat of a collapse in the "stability" of ecosystems (whatever that means)' (Walker, 1992, p. 22), Walker's own concern seems to focus on the loss of species as such. In the end, however, he appears to be caught in a hen–egg dilemma: should we care for biodiversity on behalf of the functioning (and persistence – which he distinguishes from functioning) of ecosystems, or should we care for functioning ecosystems on behalf of the preservation of biodiversity?

The Ehrlichs' book was a precursor of the intense debate on the value of biological diversity, which started to gain impetus only in the second half of the 1980s, and it was even more a precursor of the current debate on biodiversity and ecosystem functioning. In this chapter, I will briefly sketch the history (Section 3.1) and current state (Section 3.2) of this debate. My aim here is not so much to decide if, how, and to what degree biodiversity 'really' contributes to ecosystem functioning. Instead, I want to analyse how the notion of 'ecosystem functioning' itself is understood, conceptualised, and put into practice (i.e. how it is measured) in this research field. In order to better understand the current debate, it is necessary to start with its history.

3.1 A brief look at the history of the biodiversity–ecosystem functioning debate

The idea that diversity, in particular biological diversity, is not just inter-esting and/or beautiful, but that it plays an important role for the main-tenance of our world is not new. It had been – in a non-scientific form – subject of both philosophers and theologians, as well as natu-ralists. Arthur O. Lovejoy (1936 [1957]) traced the idea of the value of diversity, and of the necessary existence of the greatest possible diversity, back from antiquity through modern times. He described it as a basic metaphysical principle of Western thought. In his book, *The Great Chain of Being*, which is considered to have founded the history of ideas as a research field, Lovejoy elaborates several basic principles, which together constitute the idea of the 'great chain of being'. The first of these he called the 'principle of plenitude'. Dating back to the Greek philosopher Plato, this principle postulates that all things which are possible also must exist. According to Lovejoy, the principle of plenitude refers to: '[all] deductions from the assumption that no genuine potentiality of being can remain unfulfilled, that the extent and abundance of the creation must be as great as the possibility of existence and commensurate with the productive capacity of a "perfect" and inexhaustible Source, and that the world is the better, the more things it contains' (Lovejoy, 1957, p. 52).

That is, the principle of plenitude constituted a cosmological (and/or theological) necessity, and was also highly influential for the develop-ment of the 'research programme' of eighteenth-century biology. In a similar way, Clarence J. Glacken (1967), in his extensive study on the relations between nature and culture in Western civilisation, described this principle as an idea that is found throughout history as an important element in the thinking about humans' relations to nature. In Glacken's account it also becomes visible how boundaries between natural history, philosophy, and theology have often been difficult to draw (see especially the physico-theologians of the eighteenth century).

The question about the function and/or value of (biological) diversity thus was never purely academically descriptive, but was always embedded into cultural and religious contexts, into cosmologies, into the question about human relations with nature – and with God. Even though the current discussions on biodiversity and ecosystem functioning may not directly build on the older philosophical discourses, much of these old traditions has most likely influenced ecological theory and ecological research by means of the general cultural traditions of Western thought. Thus, in a remarkable more recent interview, one of the founders of modern conservation biology, Michael Soulé, responded to the question of Edward Grumbine, of why he thought that (bio)diversity was good:

I don't know. It's an intuition of mine and probably of many ecologists and natural historians. Whether it's because we grew up in a diverse world, or whether there's something more genetically conditioned [. . .] In other words, I love diversity. I love seeing a wide range of species and habitats. It's an aesthetic experience and it's hard to define what that is, and it's hard to define what the difference is between aesthetic and spiritual. *(Grumbine, 1994a, p. 103f.)*

This is a remarkably honest answer, and I think that many, if not most people doing research about biodiversity (myself included) share it – not denying that there are also other reasons why biodiversity should be protected.

Any attempt to reduce the current debate about the relations between biodiversity and ecosystem functioning to a purely scientific (basic science) discussion would thus miss the subject dramatically. This also becomes clear when we look at the history of the debate within ecology.

3.1.1 Two diversity debates in ecology

In ecology, there have been two debates about the role of diversity for ecological systems. The first debate, on 'diversity and stability', started in the 1950s; the second, on 'biodiversity and ecosystem functioning', was initiated in the 1980s. While the stability of communities and ecosystems, the focus of the earlier debate, is only one of several characteristics investigated in the current biodiversity–ecosystem functioning (BEF) discourse, both debates share many common questions and problems.

Two seminal publications stood at the onset of the first debate. One was Charles Elton's now classical book, *The Ecology of Invasions by Animals and Plants* (Elton, 1958), in which he formulated his idea on the relation between diversity and stability. Based on 'mathematical speculations' (Elton, 1958, p. 146) and some scattered observations from the literature, Elton tentatively hypothesised 'that the balance of relatively simple communities of plants and animals is more easily upset than that of richer ones; that is, more subject to destructive oscillations in populations, especially of animals, and more vulnerable to invasions' (Elton, 1958, p. 145). The passage is found towards the end of the book, in a chapter entitled 'The reasons for conservation'. There, he first names – very much in general – three reasons for biological conservation – namely religious (in the broadest sense), aesthetic and intellectual (including the value for recreation), and finally 'practical' reasons (concerning nutrition, health, etc.) – and describes possible conflicts between these different reasons. Elton concludes:

But suppose the conflict between these interests is not quite so great as it seems at first sight? Suppose one could make out a good case for conserving the variety of nature on all three grounds – because it is a right relation between man and living things, because it gives opportunities for richer experience, and because it tends to promote ecological stability – ecological resistance to invaders and to explosions in native populations. This would be a fourth point of view – an attempt to harmonize divergent attitudes. Unless one merely thinks that man was intended to be an all conquering and sterilizing power in the world, there must be some general basis for understanding what is best to do. This means looking for some wise principle of co-existence between man and nature, even if it has to be a modified kind of man and a modified kind of nature. This is what I understand by Conservation. *(Elton, 1958, p. 145)*

Elton thus started his statement about the relation between diversity and stability (or complexity and stability), as this discussion soon came to be known, not from a mere academic interest but from a strongly applied one – related to social, cultural, and even ethical and religious motivations.

The other 'founding publication' of the diversity–stability debate, Robert MacArthur's (1955) highly theoretical paper on the relation between species numbers and community stability was, however, of a purely academic character, with no references to any applied issues. MacArthur, by means of a mathematical treatment of food webs, also argued for a stabilising role of diversity.

Like MacArthur's paper, most of the intense diversity and stability debate that followed during the next decades (see Goodman, 1975; Trepl, 1995; McCann, 2000 for overviews) was focused not so much on conservation but on ecological theory. In contrast to the current biodiversity debate it was also a discourse led almost exclusively inside the scientific community. There was thus no need to label diversity as 'biological', because everybody knew what it referred to. As I mentioned already, the dependent variable to which diversity was related was not 'function(ing)' but 'stability'. But as we will see later, stability (or different expressions of this concept, such as persistence or resistance) is today also used as one important indicator or measure for ecosystem functioning. As another difference, the major response variables investigated were mostly properties of populations and communities, and much less biogeochemical processes of ecosystems.

Huge numbers of papers and books have been written on the relation between diversity (sometimes also referred to as complexity) and stability. While at first the major arguments were in favour of a positive correlation between these two variables (i.e. diversity begets stability), the mathematical models of Robert May (1974) contradicted this thesis. According to May, a higher ecological complexity decreased stability. The controversy about how diversity and stability were related raged for quite some time, with arguments derived from empirical as well as theoretical studies. In the late 1970s the debate ebbed because it seemed irresolvable.

In a kind of aftermath to this debate, Pimm (1984) was able to show very distinctly that an important reason why the debate appeared irresolvable, why the question resulted in so many contradicting answers, was conceptual confusion. Different measures were used for diversity or complexity (namely species richness, connectance, i.e. links between species in food webs, interaction strength, and evenness) and stability ('stable' as returning to an equilibrium point, as resilience, resistance, or variability). Also, the response variables to which 'stability' referred (individual species abundances, species composition, or total density/biomass of the species in a trophic level) differed. Sorting out the several different questions hidden behind the oversimplified question about the relation between diversity and stability led to much more consistent answers. It also revealed many gaps with respect to empirical data and theoretical work. In fact, there was not one question (or hypothesis) at stake, but several, which had been mixed up too often.

Not long after the debate about diversity and stability had ebbed, 'biodiversity' entered the stage, opening the current discourse about the protection of biodiversity and about the relevance of biodiversity for the functioning of ecosystems. This discourse started not with academic ecological questions, but was initiated and continued explicitly from a conservation perspective, namely concern about the increasing loss of species. Prominent conservation biologists, such as Paul Ehrlich, Thomas Lovejoy, Michael Soulé, and Edward O. Wilson were leading figures at the start of the new diversity debate. Although there had previously been some scattered publications on the issue (e.g. the Ehrlichs' 1981 book), the jump-start came from a conference of the US National Academy of Sciences and the Smithsonian Institution, the so-called Forum on BioDiversity, which took place in Washington, DC in 1986 (see Wilson,

1988, which is the proceedings volume of this conference). The term 'biodiversity' was also introduced to allow an easier communication to the broader public, 'biological diversity' appearing as too bulky and not appealing and/or emotional enough (see quotes from an interview with Walter G. Rosen, one of the initiators of the conference, in Takacs, 1996, p. 37).

From there on, biodiversity received a steep rise in awareness and importance, both in the scientific (see Fig. 3.2) and political realms. A culmination point of the latter was the adoption of the Convention on Biological Diversity (CBD) at the Rio Earth Summit in 1992. In the course of the negotiations for this convention (described in detail by McConnell, 1996), the context of the biodiversity issue was broadly extended beyond purely scientific and even conservation concerns, involving also economic and social aspects (see Section 5.3 for details).

In comparison to the old diversity–stability debate, biodiversity, also in its purely biological dimensions, is now in general conceived as being a much broader concept than species diversity, covering also genetic diversity and habitat (or ecosystem) diversity (see below). It is thus almost synonymous with 'life on earth'. Thus, several prominent biodiversity researchers, who David Takacs interviewed for his excellent book on the history of the biodiversity discourse, stated that for them 'biodiversity' could be defined as 'life in all its dimension and richness and manifestations' (Michael Soulé), a 'shorthand for the richness of life' (Reed Noss), or 'the living resources of the planet' (Paul Ehrlich) (Takacs, 1996, p. 46 ff.).

In the scientific realm, research on biodiversity at first focused mainly on stock-taking, on a description of the earth's biological diversity. The second half of the 1990s, however, saw rising interest in the consequences of biodiversity for ecosystem processes and for ecosystem functioning (see Figs. 3.1–3.3). This development was foreshadowed by the Ehrlichs' book from 1981, in which they referred to ecosystem functions and services as the benefits derived from species (it seems that this was the first time the term 'ecosystem services' was used). The seminal book pushing research and debate on the relations between biodiversity and ecosystem functioning, however, was issued by Ernst-Detlef Schulze and Harold Mooney in 1993. *Biodiversity and Ecosystem Function* was the outcome of a conference on the topic which took place near Bayreuth, Germany in 1991, being the kick-off conference for a programme on the issue, sponsored by several international institutions (SCOPE, IUBS, and UNESCO's Man and the Biosphere Program). Since then, the idea has become a prominent

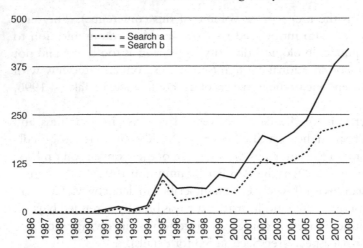

Fig. 3.1. Number of papers per year referring to biodiversity and ecosystem functioning. The results in Figs. 3.1–3.3 are derived from a search in the Web of Science (9 April 2009) using the following search routine: (a) Topic = 'biodiversity' OR 'biological diversity' AND Topic = 'ecosystem functioning' OR 'ecosystem function'; (b) the same search, but in addition to 'ecosystem functioning' OR 'ecosystem function' also 'ecosystem processes' OR 'ecosystem services' OR 'ecosystem performance' were included.

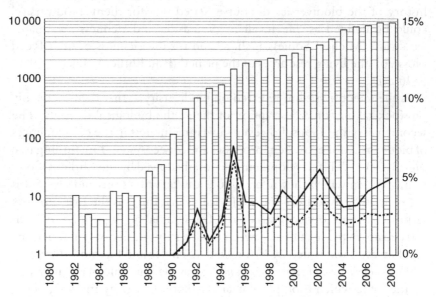

Fig. 3.2. Papers on biodiversity/biological diversity (columns, left ordinate; logarithmic scale) and percentage of those papers also referring to ecosystem function(ing) (lines, right ordinate). See Fig. 3.1 for data basis and search algorithm.

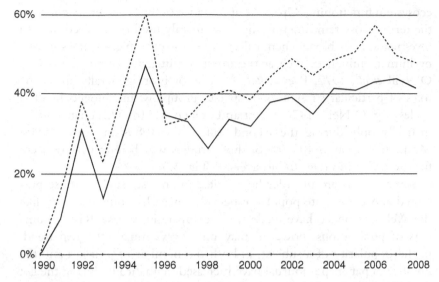

Fig. 3.3. Percentage of papers per year related to ecosystem function(ing) which also refer to biodiversity. See Fig. 3.1 for data basis and search algorithm.

field of biodiversity research, to which several books (see especially Kinzig *et al.*, 2001; Loreau *et al.*, 2002a; Naeem *et al.*, 2009a) and several hundred papers (see Fig. 3.1) were devoted.

3.1.2 Biodiversity and ecosystem functioning research: some general trends

Although, if we believe a very simple analysis of publications from the Web of Science (Figs. 3.1–3.3), only about 2.5–5% of the papers dealing with biodiversity in 2008 also referred to 'ecosystem function' or 'ecosystem functioning' (Fig. 3.2), the issue has become quite prominent and politically highly visible, more so if we add contributions relating biodiversity to 'ecosystem processes' (as one important meaning of ecosystem function), 'ecosystem performance' and 'ecosystem services' (also often subsumed under 'ecosystem functioning' – see below), the latter being an issue central to the Millennium Ecosystem Assessment (2003, 2005) and its follow-up processes.

The relation between biodiversity and ecosystem functioning is also the most dominant issue of all studies referring explicitly to ecosystem functioning. Up to the appearance of the biodiversity concept, the term 'ecosystem function(ing)' was mainly used in connection with investigations of biogeochemical cycles – for example, changes caused by human influences such as the input of nutrients or pollutants (e.g. O'Neill *et al.*, 1977; Breymeyer, 1981) – or more generally about the roles of particular substances or species groups within biogeochemical cycles (e.g. O'Neill, 1976; Grant and French, 1980). This changed dramatically only during the second half of the 1980s and early 1990s. Meanwhile, around 40–50% of those studies which refer to ecosystem function(ing) also refer to biodiversity (Fig. 3.3).

Some caveats are in order here. Citation analyses such as those presented above are quite popular, especially since electronic databases like the Web of Science have made them comparatively easy. Rising numbers of publications, however, may not always mirror the real trends in the popularity of a subject, as both the number of journals and the number of papers per journal have increased dramatically during the last decades – some journals have even increased their number of papers per year by a factor of ten. So, a strong increase might be expected for many subjects. If the number of papers per year on a particular subject are standardised with respect to the overall number of scientific papers in the respective year, sometimes different temporal trends appear. Taking this into account, steady increases in publication numbers per year may transform into a fluctuation instead – for example, if we look at the topic 'island biogeography', as we did for ten selected major journals (Heger and Jax, unpublished data). By the same analysis, however, the general trend displayed above in Fig. 3.1 remained unchanged. The trends displayed here are certainly not artefacts in this respect.

Also, databases only (can) include a limited number of journals and normally do not consider publications in books. Moreover, in our case, many studies, which today use the words 'ecosystem function' or 'ecosystem functioning' to describe some phenomena would have been carried out earlier under different headings, such as biogeochemistry. In return, some earlier studies refer to ideas captured today under the heading of ecosystem function(ing) without using this term (the same can be said for the term 'biodiversity' and other 'modern' terms in ecology). This will become evident during the following chapters of the book. But let us first look at what exactly is investigated today in the research about biodiversity and ecosystem functioning.

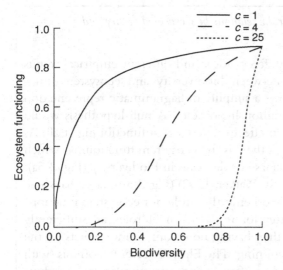

Fig. 3.4. A typical (in this case, hypothetical model-based) representation of different relations between biodiversity and ecosystem functioning. The variable c here stands for different assumed degrees of interaction between species. Reprinted from Naeem, 2002, with permission of the Ecological Society of America.

3.2 Biodiversity and ecosystem functioning: what do we measure?

3.2.1 Biodiversity and ecosystem processes

The overall question on the relation of biodiversity to ecosystem functioning, originating from a concern about the worldwide loss of species (Schulze and Mooney, 1993; Naeem, 2002), has prompted a large number of theoretical ideas as well as many empirical, increasingly experimental studies. The different hypotheses formulated about the shape of the BEF relation have been summarised and displayed by several authors (e.g. Naeem, 1998; Gindele, 1999; Schwartz *et al.*, 2000; Naeem *et al.*, 2002). In general, the studies display biological diversity as the independent variable (x-axis) and ecosystem functioning as the response variable (y-axis) (Fig. 3.4). Different shapes of this relation have been found and/or hypothesised, with Fig. 3.4 providing only a small subset of the possible options (see Box 3.1 for more details).

Box 3.1 *Hypotheses about the relations between biodiversity and ecosystem functioning*

There have been many different ideas and different empirical results regarding the relations between biodiversity and ecosystem function(ing). Figure 3.5 shows a simplified diagrammatic representation of some of the more common hypotheses. A null hypothesis would be that biodiversity has no effect on ecosystem functioning at all. As long as there are organisms there is full ecosystem functioning, which is not affected by any increase or decrease in biodiversity (Fig. 3.5a). The redundancy hypothesis (Walker, 1992) (Fig. 3.5b) states that some species can substitute each other in their roles for ecosystem functioning and that a kind of saturation is reached if all 'roles' are sufficiently filled, additional species then having no or only minor effects on the degree of ecosystem functioning. The Ehrlichs' rivet hypothesis (with which this chapter started) is often seen as very similar to the redundancy hypothesis (Ehrlich and Walker, 1998). If we take the image of the rivet popper seriously, however, it can also be interpreted quite differently (Fig. 3.5c): it assumes that many species can get lost unnoticed, but that at some point (below a threshold level of biodiversity) a complete loss of function will be experienced (or at least a rapid decline to some lower level). The rivet hypothesis has sometimes (erroneously) also been interpreted as a linear relationship between biodiversity and ecosystem functioning. A linear relationship, however, would assume that any species has an equal share in ecosystem functioning and is unique at the same time (Loreau *et al.*, 2002b). The more species are there, the higher the degree of functioning (e.g. the rate of a process like nitrogen retention). The idiosyncratic hypothesis (Lawton, 1994) states that each species is unique in its role and that it is not the number of species (or functional types) which is decisive for the degree of ecosystem functioning, but the specific biology of each species in a specific context (Fig. 3.5d). It thus denies the possibility of gaining simple predictions from species numbers alone, and instead emphasises species identity. It does not deny, however, the influence of species on ecosystem functioning as such. A variety of the idiosyncratic hypothesis is the idea that the order in which specific species are lost or added is decisive for the shape of the curve (Larsen *et al.*, 2005). Thus, as long as a keystone species is present, species number might decrease with only small or no impact on ecosystem function, but if the keystone

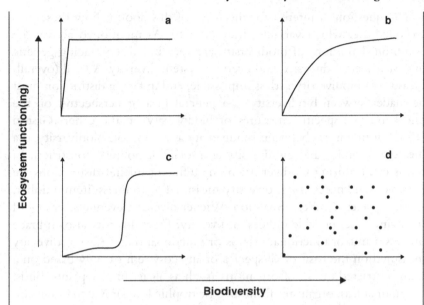

Fig. 3.5. Different hypotheses about the form of the relation between biodiversity and ecosystem functioning. (a) null hypothesis; (b) redundancy hypothesis; (c) rivet hypothesis; (d) idiosyncratic hypothesis. See text for details (adapted from Gindele, 1999, strongly modified).

species is removed, ecosystem functioning will drop rapidly, and in addition other species will soon be lost too. More recently, even an exponential relation between species diversity and ecosystem functioning was reported (Danovaro *et al.*, 2008), for which mutualistic interactions might be a possible explanation (Loreau, 2008). There are certainly also intermediate forms between the different curve shapes, and different (often combined) mechanisms can be responsible for their expression.

Please note that in general all of these hypothetical curves are normally drawn with a positive slope; in principle (and sometimes also in reality) the slope can also be negative or zero. In their analysis of 103 papers, with 446 measures of BEF relationships, Balvanera *et al.* (2006) reported various forms of the effect curve, with positive as well as negative slopes (in the sense that increasing biodiversity either increases or decreases ecosystem processes), as well as more than 100 idiosyncratic responses.

The question of interest for the topic of this book is how these axes are filled, i.e. which variables they refer to. As mentioned above, the common definitions of biodiversity are very broad, embracing genetic diversity, species diversity, and even ecosystem diversity. A total (overall) measure of biodiversity is thus impossible, and in fact, a distinction must be made between biodiversity as a general idea or perspective on the one hand and specific measures of biodiversity on the other (Gaston (1996) distinguishes between biodiversity as a concept, biodiversity as a measurable entity, and biodiversity as a social or political construct). In terms of measuring biodiversity, many different further distinctions are possible – and necessary. Is diversity measured by richness (item number, mostly species number), Shannon–Wiener diversity, evenness, or any of the many other possible indices of diversity? Does diversity also embrace processes and/or functional groups or functional traits? Also, is diversity measured on the basis of all species of an ecosystem or only based on a more restricted group of organisms (such as algae, higher plants, birds, fungi, or at least organisms from specific trophic levels)? Much discussion has been and is led about these questions (e.g. Magurran, 1988, 2004; Harper and Hawksworth, 1994; van der Maarel, 1997), and no silver bullet has or can be found.

Of the many possibilities that are given, only a restricted subset of measures is realised in BEF studies. For practical reasons, mostly only subsets of organisms can be assessed in terms of their diversity, with some groups (especially herbaceous plants) assessed much more frequently than others in empirical studies (e.g. Balvanera et al., 2006). Within these groups, the most commonly measured attribute is clearly species richness, i.e. species numbers. More sophisticated diversity indices, which also incorporate measures of (relative) abundances and which were rather prominent in the old diversity–stability debate, are quite rare here. For 446 measures of ecosystem processes and properties related to biodiversity, derived from 103 publications, Balvanera et al. (2006) found only 11 instances in which evenness was applied as a diversity measure and 19 where other diversity indices were used – in contrast to 393 cases where species richness and 23 cases where functional group richness were used. However, during the last years more refined measures have increased in importance (especially evenness; e.g. Wilsey and Potvin, 2000; Mulder et al., 2004; Valéry et al., 2009).

A very important development has also been that many authors used functional diversity instead of species (i.e. taxonomic) diversity as their measure of diversity (see below).

So what is measured as 'ecosystem functioning' on the y-axes of graphs like Fig. 3.4? In many general graphs, similar to Fig. 3.4, 'ecosystem functioning' (often also called 'ecosystem function') is substituted by 'ecosystem process' (e.g. Naeem et al., 2002, p. 5). This is what is actually measured in most instances, when the relation between biodiversity and ecosystem function(ing) is assessed. Such processes are, for example, biomass production, retention of nutrients and of water, or decomposition of litter (see overviews in Schläpfer and Schmid, 1999; Schwartz et al., 2000; Giller et al., 2004; Balvanera et al., 2006; Thompson and Starzomski, 2007). Table 3.1 shows a list of frequently used variables.

Some authors, such as Shahid Naeem, emphasise that 'ecosystem functioning refers to the biogeochemical activities of an ecosystem or the flow of materials (nutrients, water, atmospheric gases) and processing of energy' (Naeem, 1998, p. 39). A similar statement is made by Michel Loreau: '"Ecosystem functioning" is an umbrella term for the processes operating in an ecosystem, that is, the biogeochemical flow of energy and matter within and between ecosystems (e.g. primary production and nutrient cycling)' (Loreau, 2008, p. R126).

Naeem et al., (2009b), in consequence, even explicitly distinguish 'ecosystem functions' from 'biotic functions', the latter, for them, referring to species interactions.

Other authors (e.g. Moulton, 1999; Rosenfeld, 2002; Boero and Bonsdorff, 2007), however, criticise this emphasis on biogeochemical processes as too narrow and urge for the inclusion of direct biological processes related to food web interactions and life-cycles. In fact, the variables used to measure ecosystem functions are much broader than just biogeochemical ones, including such processes as herbivory, predation, leaf breakdown, pollination, bioturbation, or resistance to invasive species (e.g. Knops et al., 1999; Duffy et al., 2003; Solan et al., 2004; Balvanera et al., 2005). They also include not only processes (i.e. interactions and/or rates of change of a variable) but also, from the beginning, variables better described as ecosystem properties, such as pools of organic matter, trophic structure, and food web topology, as well as diverse descriptors of stability (mostly constancy) of some of the above variables, or stability with respect to disturbances, such as drought. Here, BEF research meets with the old debate on diversity and stability. However, the use of 'stability' – as a lesson from the failures of the earlier discourse – is much more careful today (e.g. Schläpfer and Schmidt, 1999; Loreau et al., 2002b; Griffin et al., 2009), distinguishing between different types of stability, such as resistance, resilience, or constancy (for an analysis of

Table 3.1 *A non-exhaustive list of variables described and investigated as 'ecosystem functions' in the context of BEF research. The different variables are coarsely classified here as either ecosystem properties, ecosystem processes, or ecosystem services (see text). Data are derived from various sources, especially from Vitousek and Hooper, 1993; Schwartz et al., 2000; Thompson and Starzomski, 2007; Gamfeldt et al., 2008. Note that terms designating 'ecosystem functions' in the list are not necessary mutually exclusive and are taken here as appearing in the original literature, without further scrutinising for synonymous use of terms.*

'Ecosystem function'	Type of variable	Specifications/varieties
Biomass	Property	Different species groups, trophic levels, plant parts (e.g. root biomass)
Forage crop biomass	Service	
Temporal change in biomass	Process	Different species groups, trophic levels, plant parts
Per cent cover	Property	Also: total cover; mostly referring to plant cover
Productivity	Process/service	Different species groups, trophic levels, compartments (e.g. aboveground, belowground)
Resource use	Process	Different species groups, trophic levels
CO_2 flux	Process	Also: respiration
Nutrient flux	Process	Also: nutrients cycling; various nutrients
Nitrogen retention	Process	Also: retention of other nutrients (e.g. potassium, phosphorus), similar: nitrogen use
Nutrient concentration	Property	Various nutrients in different compartments
Soil fertility	Property/service	
Disease control	Process/service	Also: pest control
Grazing	Process	Various species groups
Pollination	Process	
Extinction risk	Property	Species, functional groups
Decomposition	Process	Various substrates, e.g. leaf-litter, cellulose
Sediment stabilisation	Process	
Bioturbation	Process	
Variability	Property	Various of the above variables and additional ones (CO_2 flux, biomass, species density)
Stability	Property	Various of the above variables and additional ones (see variability), various definitions

Table 3.1 (*cont.*)

'Ecosystem function'	Type of variable	Specifications/varieties
Compositional stability	Property	Different species groups, trophic levels, different measures
Perturbation resistance	Property	Various response variables (e.g. species composition, biomass) and various perturbation types (e.g. drought, species invasions)
Invasion resistance	Property	Also: invasion susceptibility, invasion impact
Resilience	Property	Various response variables and various perturbation types
Exergy	Property	

different stability concepts and the terminological confusion in this field, see especially Pimm, 1984; Grimm and Wissel, 1997).

Still other variables investigated as ecosystem functions reach beyond purely scientific aspects of ecosystems. Increasingly, ecosystem goods and services are measured as response variables for ecosystem function(ing) (e.g. Palumbi *et al.*, 2009; Philpott *et al.*, 2009; Winfree and Kremen, 2009). Giller *et al.* (2004) even summarised the latter as 'ecosystem values', thus emphasising the normative dimension of such variables. The distinction between ecosystem processes and ecosystem services is, however, made differently by different authors (see Section 4.1).

Many BEF studies produce graphs of the kind displayed in Fig. 3.6. Results are based increasingly on experimental approaches (see Box 3.2). Some of these experiments also historically precede the current BEF debate and were made under different headings, especially in the field of agriculture and forestry (e.g. several of the studies analysed by Schläpfer and Schmidt (1999), some preceding the 1991 Bayreuth conference for more than ten years). Generalisations regarding the precise relation between biodiversity and ecosystem functions in this sense of relating to ecosystem processes and properties are, however, difficult to arrive at. Although the kind of asymptotic correlations between species richness displayed above are quite common, they are not universal. The existence and/or the precise form of the correlation between biodiversity and ecosystem functions (processes and/or properties) seems to depend very much on the specific processes or properties ('functions'), species groups, and systems considered (see also Martinez, 1996; Cardinale *et al.*, 2000; Schwartz *et al.*, 2000; Bellwood *et al.*, 2003), even if biodiversity

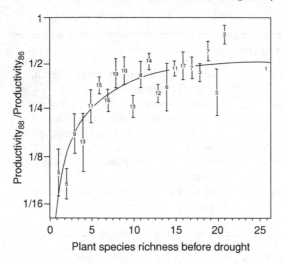

Fig. 3.6. Classical BEF response curve, derived from an experiment on the effect of plant species richness on drought resistance. Drought resistance was characterised as the ratio of productivity during drought (1986) and after drought (1988). Figure reprinted from Tilman, 1997, p. 107; see there for further details of experimental setup. From *Nature's Services*, edited by Gretchen Daily. Copyright © 1997 by Island Press, Reproduced by permission of Island Press, Washington, DC.

Box 3.2 *BEF experiments*

Biodiversity experiments are considered a major tool for disentangling the complex relations between biodiversity and ecosystem function(ing). Such experiments are, in general, laborious and require very thorough planning. Most larger experiments deal with higher plants, especially herbaceous ones. Some experiments have been done in mesocosms and glasshouses (e.g. the ECOTRON facility in the United Kingdom (Lawton *et al.*, 1993)), but most of them were performed in the field. Grassland experiments clearly dominate. The largest and most well-known ones are the Cedar Creek experiment in Minnesota (e.g. Tilman *et al.*, 1996, 2001), the BIODEPTH experiment, performed at eight sites throughout Europe (Spehn *et al.*, 2005), and the Jena experiment (Roscher *et al.*, 2004; Marquard *et al.*, 2009), in Germany. More recently, the first systematic experiments with trees were started. These range from rather small experiments with few species, such as Kreinitz in Germany (Saxonia), to larger ones with many more species included, such as the SABAH experiment in Malaysia (carried out jointly by European and Malaysian researchers),

Fig. 3.7. The BIODEPTH experimental facility at Bayreuth, Germany. Photo: Courtesy of Michael Scherer-Lorenzen, July 1998.

or the new BEF–China project (European and Chinese researchers). All in all there are currently 11 BEF experiments with tree species in seven countries, linked through the research network TreeDivNet (Harald Auge, personal communication, October 2009). Experiments were also conducted on soils and aquatic environments (Giller *et al.*, 2004; Hättenschwiler *et al.*, 2005; Stachowicz *et al.*, 2007).

The basic scheme of such experiments is rather similar and may be illustrated by the BIODEPTH experiment ('Biodiversity and ecological processes in terrestrial herbaceous ecosystems: experimental manipulations of plant communities'). In this experiment, the original vegetation was removed and the seed bank eliminated by sterilisation. Then 64 plots (2 × 2 m each) with different numbers of herbaceous plant species each were planted at each site (Fig. 3.7). Mixtures of randomly chosen species from the local species pool were sown, from monocultures through cultures with 16 species (and at one site 32) (Neßhöver, 2005; Spehn *et al.*, 2005). For most species numbers, several permutations were planted in order to avoid (or average out) any overarching effects of specific species compositions in favour of species richness as the independent variable. A special – and to my knowledge unique – feature of this experiment was the fact that it was

(cont.)

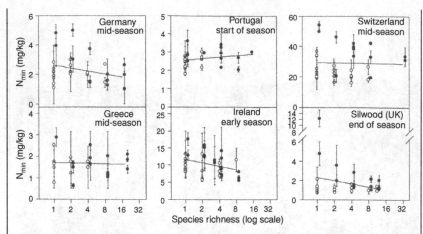

Fig. 3.8. Species richness effects on soil inorganic nitrogen concentrations in the rooting zone in year 3 of the BIODEPTH experiment. Results of the same experimental setup at sites in six different countries are displayed. Reprinted from Spehn *et al.*, 2005, with permission of the Ecological Society of America.

simultaneously conducted in different climatic and biogeographical regions of Europe, from northern Sweden to Portugal, from Ireland to Greece. The duration of the experiment was three years – even longer for two of the sites. During the three years, species composition and species richness of each individual plot were kept constant by weeding out all other species that tried to invade a plot. The dependent variables (ecosystem functions) analysed in the experiment were: vegetation cover, aboveground biomass and root biomass, canopy structure, soil nitrogen and nitrogen in aboveground biomass, and decomposition rates (Spehn *et al.*, 2005). Some of these variables were assessed each year, some only after three years.

Like in other larger experiments, the results of the experiment, i.e. the effect of biodiversity on ecosystem functions, varied depending on the specific response variables and – in this case – the specific site conditions in the different countries (e.g. soils of different initial productivity). Whereas some variables (e.g. aboveground biomass) increased rather consistently with increasing species richness (with different slopes and shapes of the response curves, however), others (e.g. soil nitrogen content (see Fig. 3.8)) showed less consistent responses, varying between sites.

While experiments like those described above (dubbed 'synthetic assemblage experiments' by Díaz *et al.* (2003)) are still the dominant experimental design, another, less frequent but increasingly important type of BEF experiment does not use artificially created communities with (often) randomly selected species combinations. Instead, these experiments start from 'natural' or 'semi-natural' ecosystems and then stepwise remove species, sometimes randomly, but mostly non-randomly (Díaz et al., 2003; Zavaleta et al., 2009). Such 'removal experiments' are seen as being more realistic. They can be used to emulate local extinction processes from established ecosystems. Removal experiments put a stronger emphasis on the effects of species or functional trait identity on ecosystem functions, on the order of species loss, and on species interactions. Both types of experiment supplement each other, being able to address different aspects of the complex of BEF questions.

There has been much discussion about the appropriate methodology and interpretation of different experiments in order to make them more realistic and to avoid methodological artefacts (e.g. Naeem, 2000; Wardle et al., 2000; Schmid et al., 2002; Leps, 2004; Schmid and Hector, 2004). Nevertheless, experiments are still seen as a cornerstone for unravelling the relations between biodiversity and ecosystem functioning, because they provide a way of at least partially controlling and disentangling the large array of possible factors governing community dynamics and ecosystem processes.

is considered just in terms of species numbers. Besides that, the specific identity and/or the order of species loss or addition has been shown to be important (Smith and Knapp, 2003; Zavaleta et al., 2009), thus emphasising the importance of species identity, beyond species numbers or abundances. Especially in the 'real world', species loss is mostly not random, as implied in the classical BEF experiments, but dependent on the susceptibility of species to specific causes of extinction. The mechanisms relevant for either results have also been much debated (Hooper et al., 2005; Hector et al., 2009), as have the experimental setup (see Box 3.2) and possible artefacts generated by it (Beierkuhnlein and Neßhöver, 2006). Not the least, it is discussed as to whether or to what degree inferences from (mostly small-scale) experiments can be relevant to the pattern and dynamics of non-experimental ('natural' and managed) ecosystems. A critical point here is also that a levelling of the response curve, with no significant further increase in ecosystem functioning, often occurs at species numbers which are below those found in natural communities. A discussion of these details is, however, beyond the scope of this book. For summaries and a (preliminary) consensus of the current state of the debate see Hooper *et al.* (2005) and papers in Loreau *et al.* (2002a) and Naeem *et al.* (2009a).

3.2.2 From ecosystem processes to overall ecosystem functioning

The BEF debate is, however, not only about the relation of biodiversity to selected ecosystem processes, properties, and services. Ecosystem functioning relates also to the whole ecosystem. In fact, although the expression 'ecosystems functioning' is frequently used simply instead of 'ecosystem functions' (in terms of ecosystem processes and properties, which are implied here), it also at least alludes to some overall function (performance, operating) of the whole ecosystem (e.g. Ruesink and Srivastava, 2001; Naeem and Wright, 2003; Giller *et al.*, 2004; Hooper *et al.*, 2005), or to those processes which are thought to be the 'critical processes' of the system (Walker, 1995; Griffiths *et al.*, 2000; see also the definition of 'ecosystem functions' by Swift *et al.*, 2004). Gamfeldt *et al.* (2008) explicitly distinguish between 'ecosystem function' and 'ecosystem functioning'. They state:

Within the field of BEF, the terms ecosystem function and ecosystem functioning are often used as synonyms; the result is some confusion regarding the meaning of these terms. With an attempt to explicitly define ecosystem functions as single processes that together constitute overall ecosystem functioning we believe that our interpretation of the effects of biodiversity change will be facilitated. *(Gamfeldt* et al.*, 2008, p. 1229)*

By qualifying 'functioning' through the word 'overall', they emphasise this distinction even more strongly. The question is how to measure 'overall functioning'. Some authors explicitly state that it is enough 'to take one of these processes and patterns as an indication of the functioning of the whole ecosystem' (Boero and Bonsdorff, 2007, p. 135), and many other papers at least imply this. Others, however, deny this possibility, e.g. Giller *et al.* (2004, p. 426), who argue that 'it is important to understand that there is no single index [and hence no single variable] that can capture the entire scope of processes and properties characterizing the overall functioning of a given ecosystem' (for a similar viewpoint, see Bremner (2008)). In fact, the problem is that there must always be a selection of response variables and of the specific kind of diversity investigated. But the intensity of one process might be increased by higher species diversity, while another might not. The same process might react differently if diversity is changed in different trophic levels (with interactions between the levels also occurring) or for different functional groups of organisms. For Gamfeldt *et al.* (2008, p. 1224), ecosystem functioning thus has to be understood as the 'joint effect of multiple ecosystem functions', for which a measure has to be found that is more than just 'the average of those functions' (Gamfeldt *et al.*, 2008, p. 1224). The solution they propose is to 'define some level of each function that we find acceptable. When one function drops beneath this level, overall ecosystem functioning is no longer sustained' (Gamfeldt *et al.*, 2008, p. 1225).

The explicit discussion of this issue (which has rarely been discussed at all) is a very valuable step in BEF research. The number of processes and properties investigated as response variables to changing biodiversity is quite broad and there are also many studies that observed several of these variables in the same experiment (see some papers listed in the review of Schläpfer and Schmid, 1999; but also see Duffy *et al.*, 2003; Hector and Bagchi, 2007). But there are clearly many studies which investigated only one response variable, and those that investigated several variables almost without exception simply juxtaposed the different results or averaged them, as criticised by Gamfeldt *et al.* (2008).

Obviously, everything depends on the question of how many and which ecosystem variables have to be selected and measured to allow reasonable statements about the functioning of the whole ecosystem – if this is really the aim of the investigation at hand, of course. That does not mean investigating 'all' ecosystem processes and properties, but at least finding some appropriate indicators for overall ecosystem functioning. The selection of variables, however, does not often appear to be very systematic when it comes to finding such indicators, with some variables – especially biomass and production – being very prominent, while others are selected only rarely.

Some researchers would say, with good reason, that talking about 'functioning' in the sense of an overall functioning of ecosystems does not make much sense at all, at least if we stick to a pure natural science perspective. They argue that an ecosystem is always functioning as long as there are organisms and processes going on between them and their environment. This opinion was expressed by several environmental scientists (not restricted to BEF research) who responded to my question about what ecosystem functioning meant and when they would consider an ecosystem to be 'destroyed'.

What is at stake here, however, is very often not just some, i.e. *any* overall performance of processes, but the (at least implicit) idea of the *proper* functioning of the ecosystem. In this vein, Naeem (1998,

p. 42) distinguished between the 'failure' of an ecosystem and its 'functioning':

From an ecosystem perspective, an ecosystem fails when it ceases to provide the services or goods demanded of it. As with machines, the ecosystem does not have to collapse. For example, if a pond shifts from a desirable state as a net carbon sink to an undesirable state as a net carbon source (say, upon entering an anoxic state after eutrophication), the pond is still a functioning ecosystem, but it has failed in the sense that it no longer provides the original service desired of it.

In a way similar to how Naeem's 'ecosystem failure' depends on defining 'desirable' or 'undesirable' properties of a system, Gamfeldt et al., in their above quote introduce a normative statement when they 'define some level of each function that we find acceptable' (Gamfeldt et al., 2008, p. 1225). This may be unavoidable, but it clearly is a statement that moves us beyond purely scientific descriptions into the normative sphere. This is even more explicit when the variables included as indicators for ecosystem functioning are 'ecosystem services'. The definition of 'ecosystem services' consciously implies a normative setting of goals in terms of human needs (including even aesthetic, cultural, or spiritual functions (e.g. de Groot et al. (2002); Millennium Ecosystem Assessment, 2003, 2005; Section 4.1)). Within this framework (set by society, not by science!), the question of how different levels of biodiversity contribute to the performance of these services can be investigated within a purely descriptive context – once the goals are set. I will come back to the important issue of the normativity of ecosystem functioning in much more detail in the following chapters.

Naaem's above statement raises also some other interesting questions. We might ask if ecosystem functioning could be conceptualised not as much as a yes–no/either–or alternative, but more as a matter of either *degrees of functioning*, or – if we look at so-called alternative stable states of ecosystems (see the case study in Section 5.1.4) – one of different *modes of functioning* (in his example an oligotrophic and a eutrophic lake).

Another crucial issue, hardly discussed in the BEF literature, is the question as to how the ecosystem and its reference states are defined. Any statements about the 'functioning' of ecosystems imply an ecosystem definition, as well as statements about a reference state (or a reference dynamic; see Box 2.2) that has to be retained over a particular period of time. In fact, this is almost never done, as the definition of an ecosystem is mostly taken for granted. But neither the definitions of the ecosystem nor of its reference state are trivial tasks. Ecosystems cannot be identified or found in nature. Instead, they must be delimited by an observer. This can be done in many different ways for the same chunk of nature, depending on the specific perspectives of interest (Allen and Hoekstra, 1992; Sagoff, 2003; Jax, 2006, 2007; Chapter 4). There is a seemingly endless number of variables that might be described within any part of nature, which one might consider to be an ecosystem, which again highlights the role of selecting the crucial variables to which 'functioning' pertains and/or by which it can be measured. There is no agreement yet, which of the many possible variables is decisive here, under which circumstances, and for which specific systems.

3.2.3 Excursus: ecosystem functioning, functional groups, and functional diversity

Classical biodiversity measures, such as species richness, are based on taxonomic diversity. They treat all species alike, no matter what their ecological differences are. When the objective of research, however, is to elucidate the relations of biodiversity to ecosystem functioning, there should be a measure of diversity that relates not just to taxonomic differences between organisms, but to what they actually do in communities and ecosystems (Petchey and Gaston, 2006). As a consequence, researchers in BEF studies increasingly relate ecosystem functioning to *functional diversity* as a measure of biodiversity with a higher explanatory power than species richness alone (Hooper *et al.*, 2002). Functional diversity relates to differences in organism properties (traits) that are related to ecosystem processes. Such traits may be: growth rate, capacity for nitrogen fixation, kinds of photosynthetic pathways, or feeding type. Some of these traits are measured directly (hard traits), some only indirectly, with less direct and explicit relations to specific ecosystem processes (soft traits) (Lavorel and Garnier, 2002). The general assumption behind the functional diversity approach is that the number and type of functional traits is crucial for the specific expression of ecosystem processes and for overall ecosystem functioning. According to this assumption, different species with different ecological properties (traits) complement each other, and species with similar ecological properties may substitute each other. Consequently, an increase in species numbers which adds a new trait (e.g. a new growth form, photosynthetic type, or feeding type) to an ecosystem should have a stronger effect on ecosystem functioning than an increase in species numbers not adding new species traits.

Functional diversity is not only of interest in its aggregated quantity as an alternative independent variable on the x-axis of BEF correlation plots, but is also of interest in its quality. This means that next to the question of how many functional types can be found in an ecosystem, there are other important questions which can help to better understand the mechanisms relating species diversity to ecosystem functioning. These are: which functional types of organism affect ecosystem processes in which way? How are the species of an ecosystem distributed over the different functional types within an ecosystem? Sorting the multitude of species into a limited number of biologically relevant types reduces ecological complexity, and at the same time increases the ecological significance of classifications. The notion of functional types and functional groups is thus an essential part of functional diversity concepts.

The idea of grouping organisms into 'functional', i.e. process-based types is not new. Two major kinds of functional group concepts must be distinguished. The first of these refers to the way in which the organism affects community or ecosystem processes; the second to how the organism responds to its environment. These two types of functional group have been called 'functional effect groups' and 'functional response groups' (e.g. Catovsky, 1998; Hooper *et al.*, 2002; Lavorel and Garnier, 2002). Only the former is 'functional' in the sense of defining different roles of organisms for the performance of a system (community, ecosystem), i.e. affecting or even determining its processes. Functional effect groups may, in the extreme, even be viewed as, all groups taken together, constructing their systems, i.e. a community or an ecosystem may be described as consisting of such groups – regardless of specific species' identities (see Section 7.1). Examples are the classical trophic types ('producers', 'consumers', and 'decomposers'), the feeding types of benthic stream organisms (Cummins, 1974), or the Eltonian niche (Elton (1927, p. 63f.), likens the niche to the 'profession' of a species). Highly abstracted functional effect types of organisms are also used in systems-theoretical approaches to ecosystems, such as Howard Odum's energy flow models (e.g. Odum, 1983).

In the other category, functional response groups, classifications are not based on the organisms' roles in building and maintaining the ecological system. Instead they refer to the strategy by which the organisms care for their own survival in the system, either in response to disturbances or under specific environmental conditions. That is, classification is focused on the survival and performance of the individual organisms themselves. Examples are the classical life form types of Raunkiaer (1934), characterising the ways in which plants survive unfavourable seasons, or the CRS strategies of Grime (1979).

Differences in the responses of organisms to environmental variability, and especially disturbances or stress, can, however, gain functional importance for a community or an ecosystem. If the organisms within one specific functional effect group (e.g. primary producers) belong to different response groups, they might react differently to a disturbance (e.g. drought or flooding). Depending on their response group, some species survive, others will go locally extinct. The loss of species from an effect group will consequently be compensated for by other species. Response diversity thus becomes functional for the ecosystem and affects its maintenance. This is the very idea of ecological redundancy as developed by Brian Walker (1992, 1995). Some authors (e.g. Hooper et al., 2002) have thus proposed viewing effect groups and response groups in a nested hierarchy when applying them to the question of how they affect ecosystem functioning, with response groups nested within effect groups.

There is not one universal one-for-all classification of functional types, not even when restricted to the question of the organisms' relevance to ecosystem functioning and the assessment of functional diversity. A wide variety of schemes exist (for overviews see papers in Smith et al., 1997; Wilson, 1999; Hooper et al., 2002; Lavorel and Garnier, 2002; Naaem and Wright, 2003). Every classification must be related to more specific questions and more specific criteria that are considered to be relevant to or decisive for ecosystem functioning. The critical question is thus how to classify organisms into such groups. What are the criteria, i.e. which traits of organisms are functionally important, which are not? How do we select them? How fine-grained should such groupings be? Are coarse trophic groups enough, or should we distinguish much more detailed modes of activity in ecological processes? Should and can we proceed in a deductive way or in an inductive way? Depending on these questions, different measures for functional diversity will be derived (for overviews, see Mason et al., 2005; Petchey and Gaston, 2006).

3.2.4 Biodiversity and ecosystem functioning: some consensus and many open questions

There is agreement that biodiversity has effects on ecosystem functioning in terms of enhancing many ecosystem processes and properties, and also has effects on the many benefits which humans derive from ecosystems (Hooper *et al.*, 2005; Naeem *et al.*, 2009b). It is highly debated, however, under what circumstances (in terms of species groups, trophic levels, ecosystem types, response variables, and biodiversity measures) this relation is expressed, in what way, which mechanisms are responsible for observed correlations (especially those found in experiments), how exactly the correlation can be described, and how general the frequently found positive correlation between biodiversity and ecosystem function might be. Even more, there is much discussion about how relevant these results, largely derived from experiments, are for real-world situations. How do (and which) species losses (locally, regionally, or globally) or species additions (exotic species!) affect the functioning of largely unmanaged ecosystems and/or managed ecosystems (especially by agriculture, forestry, and fisheries)? It is also controversial how and in which ways relevant BEF is for conservation issues (Lawler *et al.*, 2001; Srivastava and Vellend, 2005).

There is a huge (and still growing) mass of publications on BEF. Investigations increasingly are becoming more sophisticated by including multiple response variables, considering more trophic levels, performing more realistic experiments, and using more complex diversity metrics (Naeem *et al.*, 2009b; Reiss *et al.*, 2009). A real framework, however, to systematically select and integrate the many different aspects discussed under the overall question of BEF is still lacking, in spite of several attempts and postulates (e.g. Naeem *et al.*, 2002; Naeem and Wright, 2003; Giller *et al.*, 2004; Gamfeldt and Hillebrand, 2008). The BEF question, like the older question about the relation between diversity and stability, is not one question but a large bundle of more specific – although related – questions. Already the review of Schläpfer and Schmidt (1999) identified 56 possible hypotheses (according to their specific classification; many others are possible). For 20 of these hypotheses they found scientific studies. Balavanera *et al.* (2006) likewise distinguished a large bundle of different hypotheses concerning the effect of biodiversity on ecosystem functioning and services. To arrive at an integrated picture of the field – such as Naeem (2002), who described BEF as a new 'paradigm' for ecology, envisioned – we need such a general framework. Otherwise, there is the danger that many studies just produce more data of the same or a similar kind as assessed before, instead of filling in clearly defined gaps within the complex mosaic constituted by BEF.

An important step towards arriving at such a framework is to structure the many different meanings and measurement possibilities of ecosystem functioning (and biodiversity itself, of course) and to try to bring them into a logically coherent order. This involves the realisation that both component terms of 'ecosystem functioning' have to be defined and clarified much more than is usually the case. This, in turn, also involves the use of more precise language. Language is just a means, not an end in itself. But it is a necessary means if we are to be clear about the content, the meanings of terms we are really relating to. I hope that the treatment of the concept of 'ecosystem functioning' in this book can also contribute a little to building the needed framework for BEF research.

It should be mentioned that the expression 'biodiversity–ecosystem functioning research' is somewhat misleading, exaggerating what has, up to now, been done in this research field. In its current state, it would be better described as being 'species richness–ecosystem (level) phenomena research'.

3.3 Conclusions from this chapter

As we have seen, even within the somewhat restricted realm of the BEF debate, there are several different uses of 'ecosystem functioning'. As in the case study in Chapter 2, discussing the beaver on Navarino Island, there are in general two major distinct meanings of 'ecosystem function' and 'ecosystem functioning'. One refers to the operating/performance of the system as a whole, and the other refers to selected processes and other properties of the system. The latter are, in part, taken as indicators or proxies for the performance (functioning) of the whole system, but, in part, also as a measure in their own right, without the idea of an 'overall functioning' of the system involved.

What should have become clear in this chapter is also that there is no general agreement about the essential variables to be measured to assess if and/or how an ecosystem is functioning. Although some variables – such as biomass, productivity, or nutrient retention – are mentioned and measured very often, this is only a small subset of the whole number of possible variables. Not much guidance is given as to which of these variables are appropriate under which circumstances. This situation is not unique to the BEF debate, but is encountered time and again, wherever the notion of ecosystem functioning (even if under a different name) is used. Clarifying the options and providing tools for selecting useful and appropriate measurements of ecosystem functioning is the main aim of this book.

An important task for achieving this aim is finding a way through the terminological jungle that exists around the topic of ecosystem functioning. Sometimes, ecosystem processes, ecosystem properties, and ecosystem services are all summarised as either 'ecosystem functions' (Giller *et al.*, 2004) or at least as expressions of ecosystem functioning. In other papers, all three of these items are summarised as 'ecosystem properties' (Balvanera *et al.*, 2006). Yet other papers treat ecosystem processes and ecosystem functions as synonymous (but distinguish between ecosystem functions, ecosystem properties, and ecosystem services), while some authors clearly make a distinction between functions and processes (sometimes without explaining it). Still others explicitly use ecosystem functioning synonymously with ecosystem services (Winfree and Kremen, 2009). Some equate ecosystem function and ecosystem functioning, others do not. So there is ample confusion about the terminology. It is not always possible to extract the specific meanings of the terms from their respective context. In addition, there is another important manner in which 'function' is used in the BEF debate. The question at stake in the whole discourse is not only about the functions or the function(ing) of ecosystems, but also about the function that biodiversity has for whole ecosystems or for humans. Thus, one of the contributions to Schulze and Mooney (1993) was entitled: 'How many species are required for a functional ecosystem?' (Woodward, 1993). Here, function refers to a role, a purpose for the performance of the system as a whole and/or for the benefit the system creates. One of the expressions of purpose an ecosystem (can) fulfil is that of 'ecosystem services'.

The use of the words 'function' and 'functioning' is yet broader than the discussion up to now has shown. But do not misread this discussion on terminology as merely being about words! The issue is not words, but about what is really meant by the words, the ideas, the very concepts behind those words. And concepts matter. They form an important part of the fabric of theory and they constitute the background of empirical research and its applications. They influence how our work is done. It is thus necessary to analyse the different uses of 'ecosystem function' and 'ecosystem functioning' – also beyond the BEF debate – more systematically and in detail, not the least because there are also some difficult philosophical issues (and traps) involved in some of these uses. This is what the following chapter will deal with.

4 · Becoming general
What is ecosystem functioning?

'Ecosystem functioning' means different things to different people. Also – and not so surprising – there are different ways and variables by which the function(ing) of ecosystems is assessed. In this chapter I will broaden the view of the different meanings of ecosystem functioning beyond the specific context of the biodiversity–ecosystem functioning (BEF) discourse and try to distinguish and clarify the most important meanings of the concept. For this purpose I will take a close look at the two parts of ecosystem functioning, namely the ecosystem concept and the concept of function(ing). Both concepts are highly ambiguous. Understanding their semantic and philosophical intricacies is an important prerequisite for clearly conceptualising and assessing ecosystem functioning.

The most general definition of an ecosystem – what I will call its 'generic definition' in the following – defines an ecosystem simply as an assemblage of organisms together with their abiotic environment. While this definition is widely accepted, it is not very helpful when it comes to assessing ecosystem functioning in practice, for example, in BEF research, ecosystem management, or restoration ecology. More specific and concrete definitions are required. There are of course many specific definitions of what an ecosystem is, but there are also considerable differences between the definitions. These differences have important consequences for research and application. I will exemplify this by means of a case study on ecosystem management in the Greater Yellowstone Ecosystem (Section 4.3.1), and also introduce a methodology to clarify and compare the different meanings of 'ecosystem' (Section 4.3.2). The necessity of providing clear definitions of an ecosystem (the meaning of which is still often taken for granted) will become even more evident once we have a better understanding of the different meanings of the concepts of 'functioning' (Section 4.1) and the important philosophical question as to whether speaking about ecosystem functioning implies the assumption of a goal-directed system (Section 4.2).

Fig. 4.1. A beaver lodge near Millbrook, New York. Photo: K. Jax, August 1996.

For this reason, I will first analyse the different meanings of 'function' and 'functioning'.

4.1 At the heart of the problem: the meanings of 'functioning'

Let us start once again with an example that illustrates the different meanings of 'function' and 'functioning' in ecology, and questions related to it. The beaver and its beaver lodge (Fig. 4.1), no matter whether in its native range or as an exotic species, is perfect for this purpose. We know the life-cycle of the beaver and its influence on the landscape quite well. We know how an ecosystem with beavers functions and which ecosystem functions are influenced or even determined by the beaver (Section 2.1); we know which boundary conditions have to be fulfilled for the beaver to be able to build its dam; we know how it manages to build its beaver lodge and to feed. The beaver (at least in its native range) is not just an established component of the landscape, it is even a keystone species which decisively influences its environment. The beaver has the role of an 'ecosystem engineer' (Jones *et al.*, 1994). One could say that the beaver itself performs various functions: maintaining and perpetuating its own population; the provision of standing waters for other species; increasing the heterogeneity of the landscape; rejuvenating tree stands; and so on. But, would the 'same' habitat be impaired in its functioning without the beaver? The forest, which otherwise partly dies through the damming activities of the beaver, might perhaps function even better without it. Isn't the species – with respect to the forest which the beaver inundates – completely dysfunctional? Is a landscape (or its ecosystems) functioning better with or without the beaver? Which condition is desirable when we are about to manage the area? The farmer, whose land is inundated by the beaver, will evaluate the situation completely differently to the conservationist who cares about wetlands and the birds associated with them. Should the judgement about the function of the beaver within a landscape depend on the question of whether the beaver is native to the area or not?

I deliberately used the words 'function' and 'functioning' in many different ways here, all of which, however, are typical for texts in ecology and conservation biology; they may even occur together in the same text. Sometimes 'function' and 'functioning' are used interchangeably, sometimes not. So what are the major meanings of ecosystem functioning and ecosystem function, and how are they related?

4.1.1 From function to functioning

In order to clarify the meaning of 'functioning', it is useful to begin by taking a closer look at the word 'function', from which it is derived. The lexical definitions of 'function' display differing meanings. The *Oxford English Dictionary* (2009; online access 30 January 2009) declares major meanings (uses) for the noun:

1) The action of performing; discharge or performance of (something) [. . .] 2) action in general, whether physical or mental. [. . .] 3) The special kind of activity proper to anything; the mode of action by which it fulfils its purpose. Also in generalized application, esp. (Phys.) as contrasted with structure. [. . .] 4) The kind of action proper to a person as belonging to a particular class, esp. to the holder of any office; hence, the office itself, an employment, profession, calling, trade. [. . .] 6) A variable quantity regarded in its relation to one or more other variables in terms of which it may be expressed, or on the value of which its own value depends.

Two further meanings refer to religious ceremony and computer science.

Etymologically, the word 'function' derived from the Latin *functio* (performance, execution), which is also the origin of the corresponding

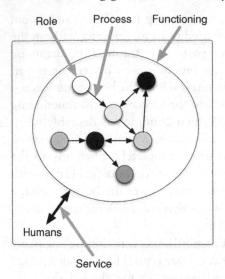

Fig. 4.2. Different meanings of 'function' in the environmental sciences (see text).

words in other European languages: German: 'Funktion'; French: 'fonc-
tion'; Spanish: 'funcion'. In the famous *Codex Justitianus*, which codified
Roman Law (around AD 530), the *functio* of a person or institution
meant its public responsibilities (its public functions), and the duties and
activities necessary to fulfil them (Thiel, 1992). As a verb, 'function',
according to the dictionary, means:

1) To fulfil a function; to perform one's duty or part; to operate; to act. [. . .] 2)
To hold a 'function' [in sense #2 above].

The term 'functioning' is directly derived from the verb. Already
the lexical and colloquial meanings of the words show the plurality of
definitions, and in principle also allude to the main senses, which we
find in ecology and other sciences.

There is a considerable literature about the meanings and (appropriate)
use of 'function' in the sciences, especially in biology (e.g. Nagel, 1961,
1977; Mahner and Bunge, 1997; McLaughlin, 2001, 2005). However,
not much has been written about the uses of these concepts within the
environmental sciences (but see Jax, 2000, 2005, from which parts of the
following have been taken). Four major uses of 'function' in ecology can
be distinguished (Fig. 4.2).

(1) As we have seen in the preceding chapters, 'function' in many cases
    is simply used as another word for *process*, or sometimes, more

specifically, *interactions*. That is, it refers to state changes in time – something that happens between two or more objects (meaning both organisms and inanimate things; shaded circles in Fig. 4.2): for example, a fox eats a mouse, or nutrients are assimilated by a plant and biomass is produced. This use of 'function' is a purely descriptive one. In many cases it also refers to the cause-and-effect relations underlying these processes. Dependent on the specific task, phenomena are, therefore, either described as temporal sequences (without considering the specific effective causes) or they are described as causal chains (see the distinction between 'pathway' and 'mechanism' in Pickett *et al.* (1987)). The use of the term 'functioning' often also refers to selected ecological processes, either single or multiple, which do not need to be correlated systematically (see also Section 3.2).

(2) In broadening the perspective a little – from a single process to several processes – a larger system, constituted of objects and their process relations, comes into view (indicated in Fig. 4.2 by the circular boundary). Here, 'function', or more often in this case 'functioning', stands for the network of processes and/or the *performance* or *operating* of the whole (eco)system. The processes as denoted by meaning (1) here are either unspecified components of the performance of the system, or they may be considered proxies, by which the performance (functioning) of the whole system is characterised. The questions by which ecosystem functioning is investigated here often refer descriptively to the different processes (meaning 2a) and are: 'What happens?'; 'Which processes occur?'; and 'How do organisms interact which each other and with their environment?'

In other cases, however, the perspective changes from a primary focus on the parts of the system to a focus on the system as a whole. This implies a different category of questions and a different meaning of 'function' (meaning 2b). The questions are now: 'How does the whole "function"?' (i.e: 'How does it operate or perform?'); 'How is the whole sustained?'; and 'What do specific parts contribute to this?'. In this more restricted meaning, the overall 'functioning' or performance of a complex system of interactions thus comes into focus, referring to some state or trajectory of the system under consideration and to the total of those processes that sustain the system. This will also be the main meaning of ecosystem function(ing) that the remainder of this book will deal with.

(3) In focusing on the relation between parts and wholes of a system, the status of the objects themselves changes. The objects characterised

as parts of the system are not mere 'protagonists' of processes, as in the first perspective, but they have become *bearers* of functions. That is: they are attributed a *role* within the system, another common meaning of 'function', which has been a predominant use of the word in its extra-scientific use. For example, a plant is seen not just as an object assimilating nutrients with the help of solar energy, but is perceived in the role of a primary producer within the ecosystem. It is possible here to distinguish between a function as such (the role, e.g. primary producer) and the bearer of the function, which is said to fulfil that function. As in a theatre, the same role (e.g. Hamlet) can be played by different actors.

This meaning of function is especially important within ecology in connection with the idea of 'functional types' and 'functional groups' (see the excursus in Section 3.2.3). On the basis of process-relevant attributes, organisms (individuals or the species they belong to) are here ordered into different ecological groups, in principle regardless of their taxonomic classification.

It should be mentioned that the meaning of 'function' as a role is not restricted to the role that something plays as part of a whole. One may also say that a worm has the function of being prey for a bird (has the role of being prey), without referring to its role within the whole ecosystem.

(4) Finally, the system is frequently extended even more by taking into consideration the relations of an ecological system with humans (the square outer boundary in Fig. 4.2). Here, 'function' is understood as something that is attributed to a system, or to selected components and processes of it, depending on its practical utility. These 'functions' mostly relate to the whole system: 'a function of the ecosystem is to provide oxygen' (for humans); or: 'an important function of streams is the elimination of sewage through self-purification'. The word 'function' here in fact denotes a particular 'service' of the system for human beings, something which fulfils human needs or purposes. This meaning is commonly implied in the context of 'ecosystem services' (see Costanza *et al.*, 1997; Daily, 1997; de Groot, 2002). In recent times, these kinds of function have become very prominent in environmental discourse and politics (in particular through the Millennium Ecosystem Assessment (MA); see the excursus below). In principle, of course, such services can likewise be described as relating to other living beings (e.g. the services a forest ecosystem provides for squirrels, or that streams provide for herons). In practice

the term 'ecosystem services' is currently used only with reference to humans.

The four different meanings of 'function' described – functions as *processes*, the *function(ing)* (*operation/performance*) of a system, functions as *roles*, and functions as *services* – are the most important ones within ecology and the environmental sciences, but by no means exhaustive. As an obvious additional meaning, function as a mathematical function should be mentioned.

In terms of their established use in the environmental sciences, 'function' and 'functioning' are not clearly separated (for examples see Chapter 3 and also the excursus below on ecosystem services). As we saw before, the words are sometimes used interchangeably, both in the meaning referring to processes (meaning 1 in the list above) and in the meaning referring to the performance of a system (meaning 2). In further discussion of the usefulness and application of the idea of ecosystem functioning, all four meanings will become relevant in one way or another. While focusing on meaning 2b (performance of a system), we will see that for many systems their (proper) performance is judged in terms of whether specific services are provided by the (eco)system (meaning 4) or whether specific functional roles (meaning 3) are realised within the system, while the processes occurring within the system (meaning 1) constitute the basis for ecosystem performance.

In what follows, I will thus not deal equally with all of the possible meanings of 'functioning'. Investigating cases in which 'functioning' only refers to some processes occurring in an ecological system is important and interesting in its own right. From both a theoretical and practical (i.e. conservation) perspective, however, it is assessing and managing ecosystem functioning in the sense of the *performance of the whole ecosystem* that constitutes a special challenge and leaves more questions open.

In the remainder of this book I will thus use 'ecosystem functioning' only in this more narrow meaning. I will use 'ecosystem processes' when talking about function/functioning in the sense of meaning 1. I will use the term 'function' only when referring to the role and importance of objects for something else, in particular for a complex system, such as an ecosystem (meaning 3). The latter is also the meaning of the term 'function' that is generally suggested in philosophical discourses (Hesse, 2000; McLaughlin, 2005).

Separating these different meanings of 'function' is not just of academic interest, but of high theoretical and practical relevance. This is due to some important philosophical differences in the character of the different concepts (and different *concepts* they are, although often designated by the same *term*). This will be investigated more in Section 4.2.

4.1.2 Excursus: ecosystem services

The idea that ecosystems provide services to humans was first discussed in the environmental sciences during the 1970s. It was dealt with under various names (e.g. 'environmental services', 'public services of the global ecosystem') both in the United States (Mooney and Ehrlich, 1997) and in central Europe, there in the context of landscape ecology (referring to 'landscape functions' or 'ecosystem functions' for society (e.g. Niemann, 1977; Haber, 1979; Vos et al., 1979)). It appears that the newer term, 'ecosystem services', was first used by Ehrlich and Ehrlich in 1981. The concept was developed more systematically during the 1980s and 1990s (sometimes also under the heading of 'environmental functions', 'functions of nature', or 'landscape functions' (de Groot, 1987; Bastian and Schreiber, 1994)). It made its grand appearance within broader circles of ecology as well as economics in 1997, when both a much cited (and much criticised) paper by Robert Costanza and co-workers, and the book *Nature's Services*, edited by Gretchen Daily, were published. A second boost, which finally brought the concept broadly into public policy, was given by the MA (2003, 2005) (see Box 4.1). Today, ecosystem services are one of the top issues of research in ecology, conservation biology, and natural resource management and policy (e.g. Kremen, 2005; Ecological Society of America, 2008; Carpenter et al., 2009).

Box 4.1 *The Millennium Ecosystem Assessment*

The MA was commissioned by United Nations Secretary General Kofi Annan in 2001. Its aim was to contribute to the Millennium Development Goals, adopted in 2000 (www.un.org/millenniumgoals). It did so by focusing on three main issues: (1) the status and trends of the world's ecosystems during the last 50 years; (2) building scenarios for future development; and (3) designing policy options. The MA is a tool for policy advice, aimed at decision makers in politics and society on different levels. Building on and integrating existing knowledge, the study included more than 1300 experts from around the world. Many international institutions and organisations were also

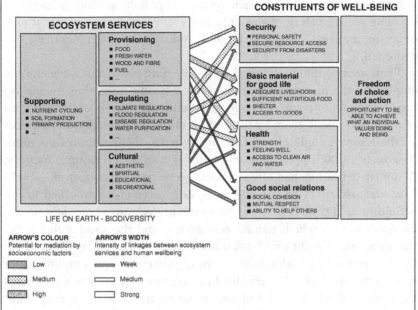

Fig. 4.3. The MA scheme showing the relation of ecosystem services to human wellbeing. The MA classified ecosystem services into four types: supporting, provisioning, regulating, and cultural; see text. Graphics: Millennium Ecosystem Assessment, 2005.

involved. The MA organised its central framework for assessing the state and trends of the world's ecosystems around the ecosystem services concept. It related ecosystem services explicitly to human wellbeing and classified the services themselves into several types (Fig. 4.3; and see below). A first volume, describing the methodological and conceptual framework of the study, was available in 2003 (Millennium Ecosystem Assessment, 2003). In 2005, six synthesis reports were published, among them an overall synthesis (Millennium Ecosystem Assessment, 2005) and five more specialised synthesis reports (see www.millenniumassessment.org for all reports and for further information on the MA). There is an ongoing follow-up process as part of the MA, aimed in particular at more detailed ecosystem assessments at sub-global scales.

Ecosystem services, by its name, is a concept linking ecology and society. The precise nature of this link, however – and with it the methodology to measure, classify, map, and evaluate ecosystem services – is contested.

There are two different directions for defining ecosystem services. One emphasises the human dimension of the concept, the other focuses on the ecological perspective. The most prominent example of the first kind of definition is that given by the MA. It characterises ecosystem services as 'the *benefits* people obtain from ecosystems' (Millennium Ecosystem Assessment, 2005, p. 26; my emphasis), i.e. as an economic category. A similar definition had already been given by Costanza *et al.* (1997). The alternative way to perceive ecosystem services is embodied in the definition of Daily (1997): 'Ecosystem services are the *conditions and processes* through which natural ecosystems, and the species that make them up, sustain and fulfill human life' (Daily, 1997, p. 3, my emphasis). Here, services are understood as those processes (and conditions) of ecosystems that *lead to* benefits for humans, but they are not the benefits themselves. While both definitions point to the same idea, they differ fundamentally in what is to be measured as an ecosystem service and how it is to be assessed.

The situation is further complicated because the use of the term 'function' confuses distinctions and connections between different concepts even more. Distinctions between services, functions, and processes are not made in a consistent way. It remains particularly unclear how much normative dimension is related to each of the terms. For Costanza *et al.* (1997, p. 253), the distinction between ecosystem services and ecosystem functions is such that 'ecosystem functions refer variously to the habitat, biological or system properties or processes of ecosystems', while 'ecosystem goods (such as food) and services (such as waste assimilation) represent the benefits human populations derive, directly or indirectly, from ecosystem functions'. However, in the table they provide, this distinction between a descriptive concept ('functions' in the sense of processes) and a normative concept ('services') is blurred. Thus 'functions' also include 'soil formation processes' as well as 'regulation of hydrological flows', and even 'providing opportunities for recreational activities'. Other authors distinguish between 'functions' as referring to the '*capacity* of natural processes and components' to provide goods and services (de Groot *et al.*, 2002, p. 394; my emphasis) on the one hand, and the services themselves, with processes underlying the functions.

Meanwhile, a thoughtful discussion on the definition and classification of ecosystem services has started, with efforts to disentangle the various descriptive and normative aspects included in the ecosystem services concept. They basically depart from the dichotomy of definitions above. Questions are whether ecosystem services are defined by (and are identical with) the *benefits* which are derived from ecosystems (as stated in the definitions by Costanza or the MA and defended by Wallace (2007)), or if services are separate from benefits, in that the latter *derive* from the services that ecosystems provide, as stated by Boyd and Banzhaf (2007) and Fisher *et al.* (2009). The latter authors explicitly perceive ecosystem services as purely biophysical (and not societal) phenomena. They thus use ecosystem services more in the sense of Daily (1997), as cited above. The debate here is not one of semantics (although some semantical aspects are of course involved), but of the usefulness of the concept for scientific and practical purposes. The specific definitions and classification of ecosystem services have major implications for the structure of research programmes about ecosystem services (as postulated by Kremen (2005) or Carpenter *et al.* (2009)), and even more for the application of the concept in different contexts. Comparisons between different sets of ecosystem services, accounting and valuation schemes for ecosystems, and management decisions all require clear definitions and classifications. This does not mean that there has to be a single fit-for-all classification. I agree with Fisher *et al.* (2009, p. 649), that the classification of ecosystem services should be made with respect to the specific context ('decision context' as they call it), and may be different when used for communication or for economic valuation. In any case, a clear account of the basic categories (benefits, services, and processes) involved is needed in order to avoid each single case being dealt with as an individual instance, prohibiting comparisons and generalisations. Otherwise, the ecosystem services concept would only be metaphorical, and only of a heuristic usefulness.

At first, I did not want to really take sides in the discussion on the precise definition of ecosystem services. As this book developed, however, it turned out that I must opt for my own working definition of ecosystem services in order to describe certain expressions of the ecosystem functioning concept (see Chapter 7). Even though the two alternative definitions point to the same idea – that humans need ecosystems and their processes and products – the differences in categories between the definitions render an overarching *and* precise definition impossible. The dilemma is: if we follow the established *economic* definition of ecosystem services (Millennium Ecosystem Assessment, 2003, 2005; Sukhdev, 2008) we have to find another word for what many call 'ecosystem services' (namely ecosystem processes such as pollination from which benefits for humans may derive). If, on the other hand, we decide to follow a definition that is closer to the *ecological* systems such as Daily's (1997) (also: IPCC, 2001; Fisher *et al.*, 2009), we clearly have to separate benefits from ecosystem services. The benefits (e.g. clean water) are then not services as such, but only the result of ecosystem services (e.g. the nitrogen cycle). A decoupling between benefits and ecosystem processes is necessary in any case – even though they are today still frequently mixed up in the same definition (e.g. Kremen, 2005; Luck *et al.*, 2009). Either ecosystem services are processes *or* they are benefits – two completely different categories of things. Nevertheless, even in the ecological definition, ecosystem services are a link to society: something *only* is/becomes an ecosystem service if it is of use to at least *some* humans.

My own working definition of 'ecosystem services' for this book is:

Ecosystem services are those components and processes which are used, required, or demanded from ecological systems (and only, if they are used, required, or demanded; otherwise they may at best be potential ecosystem services).

In a conservation context, such an approach allows us to best assess ecosystem services as an ecological category. Nevertheless, assessing ecosystem services is not something that natural scientists can do independently of societal choices: it is society that has to decide what are services and what are not. Services are thus a subset of ecosystem processes and products, depending on specific societal contexts (Fig. 4.4). Even some processes that are often taken for granted as being services, such as biomass production (leading to the benefit of food provision), are not services in any case: there may be no benefit from production or it may even be undesirable, such as algal blooms in marine or freshwater ecosystems. To assess ecosystem services in a particular region, we have to *work our way backwards* from society and its specific needs to ecosystem processes – and not vice versa, as scientists mostly do. In the context of conceptualising overall ecosystem functioning, the selection process might even go back to the point that the definition of a functioning ecosystem itself is based on the capacity of a system to provide a particular set of ecosystem services (see Chapter 7).

Note that I do not need to refer in any way to 'functions' within the scheme above. With Wallace (2007), I think we can and should skip this ambiguous term in the context of the ecosystem services discourse. Instead, it is better to talk about processes (and their products, e.g. biomass) when we refer to the ecological side of the investigation,

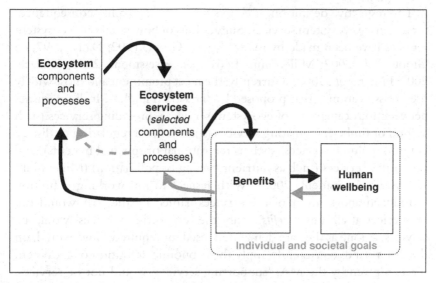

Fig. 4.4. The relation between ecosystem services, ecosystem processes, and benefits. Black arrows indicate physical effects, grey arrows indicate choices (selection activities); see text. Graphics inspired by a figure from Roy Haines-Young (in Sukhdev *et al.*, 2008).

and about *selected* (required, desired) processes and benefits when we emphasise the societal use aspect of ecosystem services.

There is much complexity also *within* the boxes labelled 'ecosystem services' and 'benefits' in Fig. 4.4, and these complexities affect the classification of ecosystem services and their assessment. Even distinguishing between services and benefits in practice is not trivial. It is beyond the scope of this book to discuss these things in detail. A few aspects should be briefly mentioned, however.

'Benefit' itself is a classical term from economics, and denotes the utility that humans gain from something. These may be of different kinds and must not necessarily be expressed in monetary terms. Benefits can be described both with reference to individuals and to society at large (or to groups of society). The precise assessment and classification of benefits is, however, controversial. Likewise, the question as to when a process (or a combination of processes and their outcomes) turns into a service, and which services contribute to produce a particular benefit, is tricky and far from trivial, especially when it comes to accounting and monetarisation (see Boyd and Banzhaf, 2007 and Fisher *et al.*, 2009 for a more detailed discussion).

The respective definition of ecosystem services also has consequences for *classifying* ecosystem services. Suggestions of how to classify ecosystem services have been made by many (e.g. de Groot, 1992; Daily, 1997; de Groot *et al.*, 2002; Millennium Ecosystem Assessment, 2003; Wallace, 2007; Fisher *et al.*, 2009). Currently, the most prominent and most widely used classification is that proposed by the MA (Fig. 4.3). It distinguishes between four categories of ecosystem services – providing services (such as food or timber), regulating services (such as flood regulation or disease control), cultural services (such as recreational or spiritual benefits), and supportive services (such as nutrient cycling or pollination) (Millennium Ecosystem Assessment, 2005). With respect to my working definition introduced above, most of the services named by the MA would not be services at all, but *benefits*. Only the 'supporting services' would be services, as long as they are demanded, used, or required. If we would, on the other hand, consistently apply the economic definition of ecosystem services given by the MA, 'supporting services' would not be services, because they are only the processes that *lead* to the provision of benefits. These distinctions may be unsatisfactory to many readers. I likewise felt somewhat uneasy when using the term 'ecosystem services' in the following text in the strict way defined above. It seems natural to speak of food as an ecosystem service instead of calling only 'the biomass production used as food' the service. However, for putting the ecosystem services concept into practice in a systematic and comparable manner, a clear decision between the 'ecological' and the 'economic' definition is necessary.

4.2 Function and functioning: teleology looming

4.2.1 Two different views on wholes and parts

Dealing with processes and causal chains can be done completely descriptively and is largely unproblematic from a theoretical and philosophical perspective. Dealing with function, functioning, and services, however, very often implies normative dimensions of different degrees. More than that, the use of function statements easily arouses the suspicion that ecosystem functioning implies an idea of ecosystems as being teleological (i.e. goal-directed) systems.

Research on ecological processes and interactions has been a major field since the beginnings of ecology as a science. It describes, for example, how individuals of different species interact in a lake or forest (or even a drop of water), which trends can be found in the long-term dynamics of the different populations or in the overall biomass of the system, or the flow of particular substances within an ecosystem. From the technical side, such a description and analysis of 'structure and function' – especially of very complex systems – is still a challenging and often technically difficult task.

New issues and questions are encountered when focusing on the performance (functioning) of the whole system. This relates to questions such as how ecosystem functioning is maintained, which roles particular species (or biodiversity) play in this functioning, and even whether there are species that may be considered 'redundant' (see Chapter 3 and Box 4.2). These questions imply some normative dimensions. Ecosystem functioning is indeed almost never a simple descriptive concept. In most of its uses, it implies not only functioning, but 'proper functioning', and thus a normative dimension. A particular state of the system is considered to be preferable to others. This is even visible in statements that deny any goal-related (teleological) connotations of the term. For example, a popular account of the relevance of biodiversity for ecosystem functioning, published by the Ecological Society of America (ESA), defines ecosystem functioning in the following way:

Ecosystem functioning reflects the collective life activities of plants, animals, and microbes and the effects these activities – feeding, growing, moving, excreting waste, etc. – have on the physical and chemical conditions of the environment. (Note that 'functioning' means 'showing activities' and does not imply that organisms perform purposeful roles in ecosystem-level processes.) A functioning ecosystem is one that exhibits biological and chemical activities characteristic for its type. *(Ecological Society of America, 1999, p. 4)*

Although any reference to the organisms' activities as being purposeful for the system (for ecosystem processes or the whole system) is denied, 'functioning ecosystem[s]' are not any ecological systems exhibiting *any* ecosystem processes, but only those 'characteristic for its type'. That is, functioning here denotes the existence of clear *reference states* of the systems and thus their 'proper functioning'.

The quote above is also a good example of the common oscillation between two phrases with a subtle but not unimportant difference: 'ecosystem functioning' and 'functioning ecosystem'. Moving from one to the other often crosses a fine invisible line. While 'ecosystem functioning' is frequently (although by no means always) merely meant as descriptively denoting some set processes occurring in an ecosystem, the 'functioning' in 'functioning ecosystem' almost always denotes *proper* functioning. Otherwise it would mean only something completely trivial and tautological: there are always some processes going on in an ecosystem, even on the basis of the most general definitions of an ecosystem.

In the same way, the common parlance that ecosystem functioning is impaired, restored, maintained, etc. alludes to the proper (or 'normal') functioning of the system. In this sense, an ecosystem – like an organism – can be said to have a goal: its continued existence. In consequence, organisms or groups of organisms can be said to have or perform *functions* (roles) for the continued functioning (performance, operating) of the ecosystem.

This brings us to a major difficulty with concepts of ecosystem functioning. While ecology, as a natural science, normally deals with questions of *how* things proceed or work, and thus with means and mechanisms, the question here is also 'what for?', and is thus related to *ends*. But ends – purposes – are normally beyond the realm of the natural sciences. The critique that arises immediately is that we are posing a *teleological* question, or even one of sense or meaning.

So, are questions about the role of a particular species within the ecosystem and about the functioning of whole ecosystems teleological?

4.2.2 Ecosystem functioning and teleology

No matter if it is intended or not, in speaking about function and functioning, teleology is always looming. If we talk about particular species or functional groups as necessary for perpetuating or maintaining an ecosystem and its functioning – a common parlance – we speak about the roles of these species at least *as if* their *purpose* was to keep the system running. In return, the functioning of a system is often envisioned as if it was goal-driven (had a goal in itself) and was determining the fate and functions of its parts.

This way of speaking alludes to organismic metaphors, viewing ecosystems as similar to 'superorganisms', as was popular in the early days of ecology. In those days, communities or other ecological systems were sometimes even literally identified with an organism (e.g. Clements, 1916; Phillips, 1935) and succession was likened to the development of such an organism:

As an organism the formation arises, grows, matures and dies. Its response to the habitat is shown in processes and functions and in structures which are the record as well as the result of these functions. Furthermore, each climax formation is able to reproduce itself, repeating with essential fidelity the stages of its development. The life-history of a formation is a complex but definite process, comparable in its chief features with the life-history of an individual plant. *(Clements, 1916, p. 3)*

This position, being the predominant way of perceiving communities for several decades (in Europe also – see Trepl, 1987; Jax, 1998), was attacked by critics like Henry Allan Gleason (1917, 1926) and others, but also by Arthur Tansley (1935), who eventually developed the concept of the 'ecosystem' as an alternative to an organismic perspective of biotic communities. One of Tansley's main points of criticism was that Clements and John Phillips (to whose writings he specifically referred in his paper) perceived of the biotic community as a 'complex organism' which as a whole (Phillips speaks of a 'holistic factor') determines the behaviour of its parts (the individual organisms, such as plants and animals). This amounts to a reversal of the classic cause-and-effect relation, an idea which violates basic principles of most philosophical and even more scientific theories (McLaughlin, 2001, p. 27). In the natural sciences, a whole cannot be the *cause* of its parts and cannot exist temporally prior to its parts.

Although the organismic approach to ecological communities (and ecosystems) became largely discredited among ecologists during the second half of the twentieth century, the organism was used, and still is used today, frequently as a metaphor for ecological systems. A prominent example of this is James Lovelock's idea of describing the whole biosphere as a superorganism, which he called Gaia (after the Greek goddess, 'Mother Earth') (Lovelock, 1979, 1989). This idea is considered highly controversial (e.g. Baerlocher, 1990; Wilkinson, 1999; Free and Barton, 2007). Another explicit attempt to 'revive' the superorganism was made by Wilson and Sober (1989), though only for a very restricted set of ecological units, not including ecosystems in general (see McIntosh, 1998 and Weil, 2005 for historical and philosophical treatments of the use of the organism as applied to ecological units beyond the individual).

As a metaphor, the organism is also present, for example, in speaking about 'ecosystem health' (Rapport, 1989; Costanza, 1992) or 'ecosystem integrity' (Karr and Dudley, 1981; de Leo and Levin, 1997; see Section 6.2), as well as in many other terms applied to ecosystems (e.g. their 'vulnerability' or 'recovery'). These more-recent approaches have also been subject to the criticism of introducing teleological (and thus 'unscientific') assumptions into ecology (Calow, 1992).

The concept of ecosystem functioning may likewise be criticised as implying teleological notions, even though these are sometimes explicitly rejected. Naeem and colleagues, in their overview of the conceptual ideas behind the current discussion on the relations between biodiversity and ecosystem functioning, state: 'By "functional" or "functioning" we mean the activities, processes or properties of ecosystems that are influenced by its biota. In no case is "purpose" inferred in our usage of these terms' (Naeem *et al.*, 2002, p. 3, footnote 2).

Nevertheless, one may argue that the very idea of the (proper) functioning of ecosystems implies some goal inherent in the system. The accusation of using teleological arguments is, however, sometimes too easily used as a knock-out argument. There are circumstances within biology for which most philosophers of science consider teleological parlance to be appropriate, if not *necessary* for scientific discourse, at least as a heuristic principle. In order to avoid both unjustified and justified allegations towards the use of the ecosystem functioning concept, it is thus necessary to deal with this subject in some more detail. I thus hope to avoid both uncritical and problematic uses of teleological propositions, as well as discarding all teleological implications completely as having no place within ecology.

4.2.3 Functions in organisms, societies, and ecosystems

The problem as such – concern about the legitimacy of 'what for?' or even 'why?' questions, and thus about teleological assumptions and explanations within biology – is not new. Particularly in connection with evolutionary theory and organismic biology, it has fostered a large number of publications (e.g. Rosenberg, 1985; Mayr, 1988a; Ruse, 1989; Pranger, 1990; McLaughlin, 2005). But as these authors also point out, such questions per se are not illegitimate for biology as a science; they are even considered as a necessary means in describing and explaining biological objects and their dynamics. However, the specific manner in which they are dealt with is crucial for the scientific rigour of biology.

Looking at ecosystems again, there are several metaphors that are applied when it comes to describing ecosystem functioning and functions in ecosystems. The most important one is (still) the organism. Another is human society (a third is the metaphor of the machine – I will briefly come back to this later). Let us look at how 'function' and 'functioning' are used in connection with these objects and what we can learn from it about the proper use of teleological statements with respect to ecosystems.

To ask about the function (the role) of the heart or the kidneys within an organism makes sense. It is also possible and important to describe the morphology and the processes of the heart without referring to its function, that is, purely in terms of physiological and even hydrodynamic characteristics – how the heart works. Here we would describe in what way and under what circumstances the muscle contracts, how the activities of the different parts of the heart interact mechanically, etc. But it is also important and even necessary to ask about the function (role) of the heart in the context of the whole organism. We would never really understand and make sense of the processes if we did not know (or at least ask for) the *purpose* of these processes – namely to provide the exchange of metabolically important substances for the whole body, the whole organism. We would not be able to differentiate between important and unimportant phenomena in connection with the heart. One might also say that the heart has the function to inspire love poems (or to produce sounds, to stay in the scientific realm), but that would not be what we mean when we speak about the function of the heart. It would be, in the words of Wright (1994), an accidental function, a 'function *as*', in the same sense that I could use my laptop or a bone 'as' a paperweight, even though nobody would claim that this was 'the' function of these objects. Talking about functions in the context of organisms (and even more with respect to their evolution) is necessary for *heuristic* reasons.

According to Kant (1790 [2009]), organisms and their parts appear to us *as if* their *existence* can only be explained by purposes inherent to them. This requires us to supplement a mechanical explanation with one referring to *final causes*. This does not suggest that such final causes are given within nature, but that our explanations of organisms would be incomplete if we would confine them only to mechanistic ones. We would not be able to pose relevant questions with respect to the organism; we would investigate an almost infinite number of secondary phenomena if we would not ask *for what purpose* the processes we investigate occur. This teleological postulate is thus for Kant a *means* to discover general underlying principles, which themselves are mechanistic cause-and-effect relations (Hesse, 2000, pp. 24ff.). 'Purpose' is thus 'not a real factor of nature which interferes with causal processes, but an idea derived by reflection, serving the assessment of a particular class of objects' (Töpfer, 2005, p. 46; my translation). It allows us to conceive of organisms as an integrated whole. A teleological perspective therefore is used as a *method* and does not refer to a different kind of causal factor.

Comparing the ecosystem with an organism means that different individual organisms (or groups of organisms) within an ecosystem are likened to parts of a body, parts which have specific functions (roles) within the (eco)system and contribute to its functioning (its wellbeing and survival, i.e. its continued performance and existence). As described above, many modern accounts related to ecosystem functioning use just this metaphor when talking about 'healthy' ecosystems or about the 'recovery' of ecosystems.

However, as I will show, it is not so much the organism (and also not the machine), but human societies that can serve as a more adequate model when it comes to describing functions within an ecological system and the functioning of the whole system.

In the social sciences there has been an influential research tradition called functionalism. This tradition, in itself very diverse, had its heyday in the first half of the twentieth century, and although highly contested (Nagel, 1961; Carlsson, 1962), is still important today, most prominently through the writings of the late Niklas Luhmann (see Moore, 1978; Münch, 2003 for overviews of functionalism in the social sciences). This line of thought describes human societies not just on the basis of the historical–causal processes, especially actions of individuals, but in terms of the functions that particular institutions, social groups, and activities display for society as a whole. There have been several schools of functionalism. From its beginnings in the nineteenth century (with Comte and Spencer), functionalists also drew on the image of an organism when describing and explaining societies and their dynamics. The philosophical problems described by e.g. Nagel (1961, pp. 520 ff.) or Carlsson (1962) for the application of function concepts to the social sciences mirror rather closely those encountered in ecology. We can thus learn a lot from the controversies in the social sciences.

Ernest Nagel cites Alfred Radcliffe-Brown, one of the major protagonists of functionalist sociological theories, for a typical expression of the perspective:

The social life of the [human] community is here defined as the *functioning* of the social structure. The *function* of any recurrent activity, such as the punishment of a crime, or a funeral ceremony, is the part it plays in the social life as a whole, and therefore the contribution it makes to the maintenance of the structural continuity. *(Radcliffe-Brown, 1952, cited in Nagel, 1961, p. 522)*

Although not as much as in ecology, there is ample confusion in what 'function' specifically means in the social science literature. Nagel (1961, pp. 522 ff.) describes six different uses of the term, the most relevant one for functionalism in the social sciences being: '[T]he function of some item signifies the contributions it makes (or is capable of making under appropriate circumstances) toward the *maintenance* of some stated characteristic or condition in a given system to which that item is assumed to belong.'

This meaning of function is also mirrored in the above quote from Radcliffe-Brown.

So, using the image of a human society, we may then say that a specific organism or process fulfils specific functions (roles) within the (biotic) community or the ecosystem (equalling society), and by these activities maintains the functioning (maintenance of structural continuity of the society) of the community or ecosystem. It is also this metaphor that Charles Elton was using when introducing his concept of an ecological niche: 'When an ecologist says "there goes a badger" he should include in his thoughts some definite idea of the animal's place in the community to which it belongs, just as if he had said "there goes the vicar"' (Elton, 1927, p. 64).

Comparing ecosystems to human society so far does not sound much different from the comparison between the ecosystem and an organism. So why should society be a better metaphor for an ecosystem than the organism? Let's look closer.

In contrast to the parts of an organism, a particular organism or species has no clearly defined role within an ecosystem: a bird may have the function of being prey to other animals – but only if these carnivorous animals are parts of the specific system. If there are no predators in the system, the same species or even individual will not have the role of 'prey'. Even if we can say that the bird actually has the role of being prey, we can also find other roles, such as its role of distributing seeds and nutrients, preying on insects, etc. That is, like a human individual within a human society, who may be a teacher, spouse, child, politician, etc., either at the same time or at different times, it can have several roles. Roles can change and the same person, as well as the same species, can even take opposing roles in time (an extreme example is given by the reversal of predator–prey relationships (Barkai and McQuaid, 1988)). 'The' one and only role of a species does not exist. Roles are strongly context-dependent. Also, species (and even single individuals of a species) can live in different ecological systems (migrating birds and many ubiquitous species) in the same way that

Box 4.2 *Can species be redundant?*

The question of whether and to what degree a species can be *functionally redundant* in ecosystems, discussed in the context of the BEF debate (e.g. Lawton and Brown, 1994; Naeem, 1998; Loreau, 2004; see also Chapter 3), receives an answer in the context of the current chapter. In order to determine whether species are 'redundant', it is necessary to precisely explicate to which processes ('functions') redundancy refers. Only then is it possible to judge if a particular species is redundant with respect to this specific ecosystem process. That is, if the role it performs in the ecosystem, such as the process of decomposition, can also be fulfilled by other species present in the system.

Walker (1992, 1995) lists a number of functions for describing ecosystem functioning, which he sees as 'critical processes (functions) that determine and maintain it' (Walker, 1995, p. 749). He also emphasises that species that are redundant at one time might not be at another, such as when climate or other abiotic conditions change (the 'insurance hypothesis' of Yachi and Loreau (1999) being a corollary of this). Beyond that, a species can be redundant with respect to one process but unique with respect to another (see Hector and Bagchi, 2007 for empirical support). 'Absolute' or 'general' redundancy of a species thus can never be determined. Everything depends on the selection of the variables to be investigated, the specific concept of the ecosystem at hand, the specific environmental setting, and – more generally – on the aim of the specific study (see also the critique of the redundancy concept by Gitay *et al.*, 1996).

people can change the society in which they live, for example, by emigrating or by living on two continents. In contrast, the heart can never take the role of the kidney, and only by means of surgery can it leave its original system.

An additional and crucial problem in describing the function of a species or particular variables (e.g. biodiversity) in or of the ecosystem is the definition of the system itself. This is even more important when the 'functioning' of the whole system shall be described. While the organism 'system' and its reference state (healthy) are given and known rather well, this is neither the case for human societies or for ecological systems. It can also easily be decided if an organism is alive or dead, while it is enormously difficult (sometimes impossible) to decide for a human society when it ceases to exist as such – when it is 'dead' (Nagel, 1961, pp. 527 ff.; Carlsson, 1962). Did the French society of the eighteenth century 'die' through the revolution of 1789? Did every French republic constitute a new society? As Nagel puts it:

[I]n regard to the condition of survival by a society, there is nothing comparable in its domain to the generally acknowledged 'vital function' of biology as defining attributes of living organisms. Societies do not literally die, though to be sure a society may disappear because all human beings who constitute it die without leaving heirs or are permanently dispersed. It is therefore not easy to fix upon a criterion of social survival that can have fruitful uses and not be purely arbitrary. *(Nagel, 1961, p. 527)*

Likewise, it is far from obvious (and will be judged completely differently by different observers and for different purposes) when and if an ecosystem has 'died', 'collapsed' or become another ecosystem (Jax *et al.*, 1998; also, see below).

Organisms set their own boundaries and reference states, their teleology is based on *internal* purposiveness (Hesse, 2000). In contrast to organisms, ecological systems are strongly dependent on the specific perspectives of the observer. The only teleological description we can apply to them without problems is thus an *external* purposiveness, and thus an external teleology – that is, one in which purposes are set by the observers, but not by the system itself.

It should be noted that there have been attempts in organismic biology to express seemingly teleological processes in other terms. Ernst Mayr (1988a) coined the word 'teleonomy' for processes following a (genetic) programme. Others rephrased some phenomena hitherto described in a teleological language as 'adaptive' in an evolutionary sense (Ruse, 1989). Both approaches have also been discussed in connection with ecosystems (e.g. Jørgensen *et al.*, 1992; Levin, 1998). However, ecosystems neither possess an inherent programme comparable to the genetic code of organisms, nor do they compete and adapt. The now rather common phrase of ecosystems as being 'adaptive' remains rather elusive, because it is not clear what – besides the individual organisms – might adapt to changing environmental conditions. As Levin (2005, p. 1077) states, it is certainly not the ecosystem as a whole that is a unit of selection. But if 'adaptive' only means that the *organisms* in an ecosystem react to these changes, and that they and their interactions are subject to evolution, the parlance of ecosystems as '(complex) adaptive systems' is a rather trivial and misleading expression – misleading because the term suggests that the whole system 'adapts'. Species in an ecosystem certainly influence each other – including in a way that has evolutionary consequences – but they are not linked to each other as closely as the parts of an organism are. There is plenty of paleoecological evidence that, for example, in the course of climate changes, species do not migrate and change as closely linked groups, but as individual species (Miles, 1987) – notwithstanding of course the joint migration of symbiotic species.

Teleological statements about ecosystems and their functioning thus must not refer to an internal teleology, as is legitimate when describing organisms. If we speak in a teleological manner about ecosystems we have to be clear that their purposiveness is only an *external* one, set by individual observers or by societal choices. Everything else can just be a very loose analogy, which often may lead more to confusion and misunderstandings than to sound science and good communication of empirical findings. The external purpose does not need to be a purpose for us (humans). We can also ask what purposes (functions) specific species have for the continued performance of a system. This purpose, however, is one depending on us selecting the system and its reference state and not a purpose which nature, or the ecosystem, brings about. *We* are *interested in* under which conditions a particular configuration of objects (e.g. species, species groups, traits) and some of their interactions persist.

A third metaphor used to describe ecosystems (and nature in general) is that of the *machine*. Especially in connection with the rise of systems theory and modelling in ecology, this metaphor has become popular (Golley, 1993; Keller, 2005). Unlike organisms, machines are clearly constructed and receive their purposes and goals from an external agent. This creates the question of who really 'constructs' the ecosystems, at least as long as non-highly anthropogenic ones are concerned. Howard Odum, particularly, pioneered this perspective of ecosystems (Taylor, 1988). He also coined the term 'ecosystem engineering' (Gattie *et al.*, 2003), a term which obviously rests on the machine metaphor. In a review of the state and history of ecological engineering, the metaphor is spelt out explicitly: 'The prototype machines for ecological engineers are the ecosystems of the world' (Mitsch and Jørgensen, 2003, p. 372). However, the boundaries between the organism metaphor and that of the machine, as applied to ecosystems, often become blurred. Ecosystems are sometimes seen as a special kind of machine that is 'self-organised' (Keller, 2005; Voigt, 2009), whatever the 'self' may mean here. Also, the idea of extracting 'services' from ecosystems alludes to the machine metaphor.

The fact that ecosystems are not given by nature as naturally delimited objects, but are strongly dependent on the perspective of the observer, makes the application of function statements to ecosystems even more problematic – and distinct from their application to individual organisms. In the social sciences, (neo-)functionalism (e.g. in the writings of Niklas Luhmann) has likewise moved away from a perspective that investigates ontologically 'pre-given' objects (societies) to an investigation of *systems* whose construction is strongly determined by an observer (Jetkowtiz and Stark, 2003).

There is thus no clarity in the determination of the *object* 'ecosystem' (a problem which we also find in defining human societies; (e.g. Carlsson, 1962)). No generally accepted definitions of the 'ecosystem' or the 'community' exist, that at the same time could be applied for providing clear reference states, describing ecological functions, or allowing statements about the 'functioning' of the system (Shrader-Frechette and McCoy, 1993; Jax, 2006). For some scientists, an ecosystem is created by all the organisms in a given area with their interactions, while others consider only those systems which also exhibit self-regulation or clear trophic structures to be ecosystems. For some, ecosystem boundaries are any boundaries defined by an observer; for others, ecosystem boundaries are given by gradients in process rates ('functional boundaries'). The answer to the question of what a *functioning* ecosystem is, is thus in part dependent on the concept of the ecosystem that is implied. The way in which we define an ecosystem, however, is partly a matter of normative settings, world views, and even societal decisions.

4.3 The meanings of 'ecosystem'

The most accepted meaning of 'ecosystem' is that of an assemblage of (interacting) organisms together with their abiotic environment – what I have called the generic ecosystem definition at the beginning of this chapter. While this general definition is broadly applicable, it is not of much use when it comes to assessing whether (and/or to what degree) an ecosystem is functioning. There is, for example, no way to count the ecosystems in an area on the basis of such a definition. What then, would be *no* ecosystem, given the existence of organisms in a place? Also, there are in fact often many implicit assumptions about the 'nature' of an ecosystem beyond what is expressed in such simple definitions.

Nevertheless, the meaning of 'ecosystem' is mostly taken for granted. But taking the meaning of this basic ecological concept as given or as in no need of precision can lead to serious problems when it comes to managing ecosystems and preserving their functioning. An excellent example to illustrate this is the controversy about wildlife management in Yellowstone National Park, not the least because the issue was framed in terms of ecosystem management very early on. The case study also demonstrates the variety of factors that influence judgements about what a functioning or intact ecosystem is, beyond purely academic discourses, not least the diverse views on what counts as 'natural'. Having started with this case study I will then move on to a more general description of how to unambiguously characterise the different meanings of the term 'ecosystem'.

4.3.1 Case study: destroying or maintaining a functioning ecosystem? Wildlife management in Yellowstone National Park

Yellowstone National Park is the oldest national park in the world. Founded in 1872, the large park in western America has been the prototype for most national parks in the world and can legitimately be called an icon of American and world natural heritage. Nevertheless, the question of the proper management of the park has been a subject of controversy from the very beginning. This relates to almost every conceivable issue in conservation, be it biological, geological, or cultural. These controversies have often raged for a long time and become very heated. Because of the importance of the area as a flagship of conservation, the debates have often influenced how the questions posed in Yellowstone were answered in other protected areas. It is therefore no wonder that Yellowstone National Park and its history are amazingly well documented and provide an interesting case study and 'experiment' for many burning questions within ecology and conservation biology.

In this chapter I will focus on the conflict about ungulate management on Yellowstone's northern range, which is intimately related to the theme of this book. As I will come back to Yellowstone as an example in other contexts, it is necessary to introduce the park and its history in more detail.

The setting

Yellowstone National Park covers an area of almost 9000 km^2 and is situated in the northwest corner of Wyoming, with small portions extending into Montana and Idaho (Fig. 4.5). At an average altitude of some 2300 m, with some mountains rising up to 3400 m, the vegetation consists mostly of coniferous forests, but also contains large natural grassland areas. It also comprises several rivers and lakes, among them Yellowstone lake, with an area of 352 km². The park harbours more than 50 species of mammals, among them large herds of ungulates, in particular elk (wapiti, *Cervus elaphus*) and bison (*Bison bison*), as well as several large predator species like bears (*Ursus arctos* and *Ursus americanus*) and grey wolves (*Canis lupus*). Yellowstone is surrounded by huge tracts of other public land, including one national park (Grand Teton National Park) and several national forests and wilderness areas, comprising together about 44 500 km² − an area larger than Switzerland. This area is commonly called the Greater Yellowstone Area or the Greater Yellowstone Ecosystem (Fig. 4.5).

The area the park covers was never settled by Europeans and was only touched and 'discovered' by a few trappers. Native Americans had been using the area to some degree, but due to its harsh winter climate they had, as far as we know, very few permanent settlements. The only tribe with permanent settlements in what is now Yellowstone National Park was the one known as 'Sheepeaters'. During the formative years of the park, these people were 'removed' from the area (see Nabakov and Loendorf, 2002).

The main reasons for establishing the park in 1872 were the scenic beauty and the many geothermal features (geysers and hot springs) of the area. Only later did the abundant wildlife become valued as another feature requiring protection and management. The management of Yellowstone has a very complex history, with many changes over time (Chase, 1987; Schullery, 1997; Pritchard, 1999). In the beginning (the 1880s) the park was protected by the US Army. In 1916 the National Park Service (NPS), as an agency of the US Ministry of the Interior, was founded. Since its founding, the NPS has been responsible for the administration and management of the park. In the early twentieth century, for example, 'good animals' like elk and bison were nurtured, while 'bad' animals like predators were combated, leading to the eradication of the wolves from the Yellowstone area (prior to their successful reintroduction to the park in 1995, the last wolves were reported in the mid 1930s).

The management of Yellowstone National Park
The management goals of Yellowstone National Park are determined by a hierarchy of influences which constrain the options of its management.

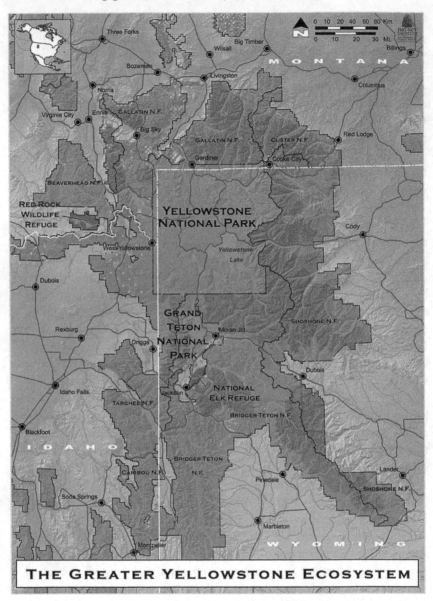

Fig. 4.5. Map of Yellowstone National Park and the Greater Yellowstone Ecosystem. Copyright: The Big Sky Institute, Montana State University. Printed with kind permission.

This starts with the legal framework pertaining to the park. The most important ones are the founding act of the park (1872) and the act that created the NPS in 1916. The latter specified the goals of the national parks as: 'To conserve the scenery and the natural and historic objects and the wild life therein and to provide for the enjoyment of the same in such manner and by such means as will leave them unimpaired for the enjoyment of future generations' (US Congress, National Park Organic Act of August 25, 1916, 16 USC §1).

This is a very general goal and it requires interpretation to be transferred into specific management objectives. Building on these juridical foundations, the NPS has issued general management guidelines, valid for the whole National Park System. It is within the framework given by these guidelines that each park develops its specific management plans on several timescales.

Up to the middle of the twentieth century, management of US national parks was mostly single-species or single-feature management and not much oriented by ecological insights (Sellars, 1997). In the 1960s, however, a major change in management strategies occurred. This change was closely connected to the rise of the ecosystem concept as a tool for the management of 'natural resources'. Several papers and at least two theses dealing with – and demanding the application of – the ecosystem concept to the management of national parks appeared in the late 1960s and early 1970s (Schultz, 1967; Barbee, 1968; McClelland, 1968; Reid, 1968; Houston, 1971). This change can be attributed to two major causes. One cause is to be seen in the general development of scientific ecology in the United States, namely the rise of ecosystem research as an accepted and institutionalised speciality, which occurred during the 1960s (Golley, 1993). The second major trigger for the shift to the management of 'wholes' or ecosystems was a report commissioned by Secretary of the Interior Stewart Udall, aimed at reviewing the wildlife policies of the NPS. The report, which was published in 1963, is now widely known as the *Leopold Report* (Leopold *et al.*, 1963). Although the word 'ecosystem' does not appear at all in this paper, the *Leopold Report* is taken by most people as the starting point of the idea of ecosystem management in the National Park System. This interpretation is based mostly on passages of the report which urge for the protection of habitats and their dynamics as a means of protecting wildlife, and especially on the postulate that the goal of national parks should not be just the protection of individual species, but of 'whole associations'. This is expressed in the most widely cited paragraph of the report:

As a primary goal, we would recommend that the biotic associations within each park be maintained, or where necessary recreated, as nearly as possible in the condition that prevailed when the area was first visited by the white man. A national park should represent a vignette of primitive America. *(Leopold et al., 1963, p. 32)*

At the same time they state:

Above all other policies, the maintenance of naturalness should prevail. *(Leopold et al., 1963, p. 35)*

The *Leopold Report* was the explicit basis for the first comprehensive management guidelines of the NPS for natural resources which were published in 1968 (National Park Service, 1968), and which widely reiterate the recommendations of that report. It thus gave a more refined and extended interpretation of the goals of national parks as described in the acts of 1872 and 1916. Here the ecosystem concept was introduced into the official management policies of the NPS. While the 'ecosystem' was only sparsely mentioned in 1968, its use and specification became extended and common in the greatly enlarged updates of the management guidelines in 1978 and 1988 (National Park Service, 1978, 1988, and beyond (Latest update: National Park Service, 2006)).

As with the general policies of the NPS, the ecosystem concept was first used in Yellowstone National Park explicitly in the late 1960s. To my knowledge this is the starting point for any use of the ecosystem concept in the management of protected areas. The concept was introduced by Yellowstone's (at the time) new research biologists, Glen Cole (1969) and – a few years later – Douglas Houston (1971), in connection with the management of elk (wapiti) populations in the park and was related to what they called the politics of 'natural regulation' (see also Huff and Varley, 1999). The management concept reached an important level with the notion of the Greater Yellowstone Ecosystem as the decisive management unit, which was proposed in the late 1970s and was soon embraced both by the managers of Yellowstone and by environmentalist groups (see Clark and Zaunbrecher, 1987; Glick *et al.*, 1991; Keiter and Boyce, 1991). In spite of opposition by some regional groups, ecosystem management and the notion of the Greater Yellowstone Ecosystem are now default parts of Yellowstone's management strategy. More than that, the goal of maintaining ecosystems (and their functioning) is generally accepted as a major goal of NPS policy, even by its critics: 'Hence, this goal of preserving intact ecosystems appears well established, and is, in fact, the central goal of natural-resources management in the [National Park] System' (Wagner et al., 1995, p. 16).

Yellowstone's northern range: is it an intact ecosystem?
There is thus a general consensus that the main goal of the management of US national parks is to maintain 'healthy' or 'intact' ecosystems (e.g. Greater Yellowstone Coordinating Committee, 1990; Wagner et al., 1995). In spite of the agreement that ecosystem management is central to the management of national parks, however, much disagreement exists in regard to the assessment of the state of the Yellowstone ecosystem and the appropriateness of the management strategies of the NPS. This strategy and its goals have been characterised by the managers of Yellowstone themselves as: 'To preserve the natural and cultural resources of Yellowstone and to allow natural processes and interactions between resources to occur with a minimum of human influence' (Yellowstone National Park, 1995, p. 2), an expression that reflects similar statements in the general guidelines of the NPS.

While the NPS, several scientists, and, especially, the environmentalist groups argued that the Yellowstone ecosystem, despite many threats and human influences, was still 'one of the largest, relatively intact temperate zone ecosystems left on earth' (Glick et al., 1991, p. 9), the critics saw the park and its ecosystem as being threatened, degraded, even destroyed by the management policies of the NPS (in particular, Chase, 1987; Wagner and Kay, 1993; Wagner et al., 1995; Kay, 1995, 1997). This discussion focused particularly on the fate of the so-called 'northern range' of Yellowstone Park (Fig. 4.6). This is an area with large grasslands at a rather low altitude (although still around 2000 m), which extends in part beyond the northern boundary of the park into the Paradise Valley of the Yellowstone River. For large herds of ungulates (elk, bison, pronghorn) it serves as their foraging area, especially during winter (see Despain et al., 1986; Singer, 1996; and Yellowstone National Park, 1997 for detailed descriptions). The critics of the park management considered this area to be heavily overgrazed and degrading due to the reluctance of the NPS to control the elk population. A large number of scientific investigations were carried out, collecting an enormous amount of empirical data to refute or support this claim (e.g. papers in Singer, 1996). They investigated patterns and dynamics on the northern range and assessed how conditions of vegetation, animal abundances, climate, human impacts, and other relevant variables had historically been in the park, especially in its early years, as a benchmark for recent conditions. But in addition

Fig. 4.6. Yellowstone's northern range: a large grassland and savannah area. Photo: K. Jax, August 1998.

to controversial empirical questions, there are in fact severe conceptual and political issues that explain why the controversy lasted so long. The analysis of these issues will help us to understand the importance of different (implicit) definitions of the ecosystem and the values which influence their selection.

The controversy about Yellowstone's northern range had its peak during the 1990s. It even led to a review of the issue, conducted by a study group of the US National Research Council (NRC). The review was initiated by the US Congress and published in 2002 (National Research Council, 2002). Meanwhile, the discussion has somewhat ebbed, for which several reasons may be responsible, such as the altered behaviour of the elk population after the wolf population (reintroduced in 1995) increased (Ripple and Beschta, 2004, 2007). An extended summary of the (fundamental) critique of the management of Yellowstone National Park was given in 2006 by Frederic Wagner (with 'replies' given in reviews of the book, such as Turner (2008) and Cooper (2008)).

So, what in fact constitutes an 'intact' or 'healthy' – we may also say 'functioning' – ecosystem, and at what point is the 'integrity' of the ecosystem destroyed? What are the precise objects to be protected or perpetuated? Several different criteria are applied in the discussion for or

Fig. 4.7. Elk (*Cervus elaphus*) near Mammoth Hot Springs, Yellowstone National Park. Photo: K. Jax, August 1998.

against the intactness of the ecosystems, some of them of a more directly philosophical and socially determined kind, some of them based on the specific theoretical perceptions of the ecosystem concept.

Naturalness One common answer to the question of what constitutes an intact ecosystem is to equate the intactness of ecosystems with their naturalness. This is a criterion for the management of the national parks and is also a well-established and explicit goal of NPS management strategies in the tradition of the *Leopold Report*. It shifts, however, the question of defining 'intactness' only to the task of defining 'naturalness' and – when applied to particular areas – to determining the 'natural' state of the ecosystem. Although almost all publications dealing with the management of national parks in general and with Yellowstone in particular argue about their naturalness, only very few provide an explicit definition as to what 'naturalness' means and how it might be assessed (exceptions are: Greater Yellowstone Coordinating Committee, 1990; Rolston, 1990; Wagner and Kay, 1993; and – to a certain degree – Wagner *et al.*, 1995). To most people, especially those from the environmentalist side, naturalness is given if no – or a minimum of – human intervention in the area is present (see also National Park Service, 1988). From this perspective, naturalness can be achieved by a hands-off policy of natural resource management. For others, including the critics of such policies, naturalness is defined by the absence of 'modern man'. These authors argue that the Yellowstone ecosystem is not in its natural state because it is missing at least one important component which is considered to have had great influence on it: Native Americans (Chase, 1987; Kay, 1995; Wagner and Kay, 1993). On the other hand, the argument can be made that, due to the presence of Native Americans (i.e. humans) in the area for a long time, Yellowstone was not in a natural state at all after it was designated as a national park in 1872. The argument can, in principle, be used either way, depending on the understanding of what is natural, and especially of the time at which humans departed from nature. It can be used to argue for a non-intervention policy that even allows for radical change, because the natural state might not be the one that was encountered in 1872 and will only be restored by really letting nature take its course; or it may be applied to justify an interventionist strategy which substitutes the 'natural' activities of Native Americans through management activities by the NPS – in particular, prescribed burning and hunting of large ungulates.

The concept of naturalness that is applied, which is in fact a philosophical and not a scientific question, thus has great importance for the direction of management in national parks (see also Pritchard, 1999). In other contexts naturalness is also frequently assumed as a criterion for 'intact' or 'functioning' ecosystems (e.g. for ecosystem integrity; Section 6.2). The influence of (mostly implicit) concepts of the 'natural' within scientific ecology is of much higher importance than generally appreciated by scientists themselves.

Historical conditions: a vignette of primitive America   Another way to define what are 'healthy' or 'intact' ecosystems in national parks is to equate this status to the conditions that were present when the first Europeans came to the areas in question. This corresponds to what the *Leopold Report* terms a 'vignette of primitive America'. Sometimes this is equated with 'natural', as discussed above, sometimes it is decoupled, to avoid the problems inherent in any definition of naturalness (e.g. Wagner *et al.*, 1995). Wagner and colleagues, in their review of wildlife policies in the national parks, stated:

Thus there is a lingering and pervasive assumption that the pre-Columbian condition, imperfectly as it is known, was a desirable one that land management, including that of national parks, should try to emulate in some degree. That condition is commonly presumed to have been in some significant degree what today we call healthy or intact. We have in previous chapters implied that same synonymy. *(Wagner et al., 1995, p. 172)*

However, it is problematic to equate a particular historical condition of natural areas with their health and intactness. There may be very different theoretical assumptions of the specific concept of 'ecosystems' in considering pre-Columbian nature to consist of 'intact' ecosystems, which must be specified (see below). Determining the way an ecosystem was at any time in the past does not tell us how it should be. Depending on both the definition of what exactly constitutes an ecosystem and on the controversial factual questions as to how big the influence of Native Americans in each particular area really was, the goals of keeping the area in a state it exhibited when Europeans first 'visited' the area and that of 'providing the American people with the opportunity to enjoy and benefit from natural environments evolving through natural processes minimally influenced by human action' (National Park Service, 1988, p. 4:1) may seriously collide.

Thus, looking for a particular historical condition as a point of reference for 'intact' and 'healthy' ecosystems neither substitutes for the discussion about conflicting values of conservation nor provides clarity when discussing the criteria used to define and specify ecosystems in general or that of particular parks.

Defining and specifying the ecosystem As mentioned above, ecosystems and other ecological units such as populations and communities can be defined in various ways. They are never just simple representations of 'reality', but by necessity are always abstractions by observers, which will be different according to different tasks and interests. A problem is that many of the definitions are not explicit, at least not in a satisfying manner. If explicit definitions are given at all, they are mostly of the following kind:

Ecosystem – Living organisms (biotic) together with their nonliving environment (abiotic) forming an interactive system inhabiting a defined area of interest. *(Greater Yellowstone Coordinating Committee, 1990, p. G-2)*

This kind of definition is rather common (similar to, e.g.: 'Ecosystem: all the organisms in a given place in interaction with their nonliving environment' (Forman and Godron, 1986, p. 592)). They come close to what I have called the generic (or minimum) definition of 'ecosystem' (see also Jax, 2002a, 2006), i.e. the lowest common denominator of almost all other definitions of this term.

Definitions of this kind, however, are too vague to assess the question of at what point the ecosystem is 'intact' or 'destroyed', as they apply to almost every part of nature. To get more detailed information on what is meant by an ecosystem (and I am sure that most scientists implicitly have a more precise idea of what an ecosystem is to them), it is sometimes necessary to use an interpretative (hermeneutic) method in the analysis of statements about ecosystem management. This method is currently not very common in the sciences, but is used more in the realm of the humanities. It does, however, provide valuable and sometimes indispensable information on issues in ecology and conservation.

In what follows I analyse the views of both the supporters and critics of the current Yellowstone National Park ecosystem management policies in regard to what constitutes an intact ecosystem for the different sides. There are several guiding questions we may use to get a more detailed picture of what an ecosystem is meant to be. The first of these pertains to how closely *integrated* a system of organisms and their environment has to be in order to call it an (intact) ecosystem.

How 'integrated' is an ecosystem?

For most of the critics, but also for some of the environmental groups who support the management policies of Yellowstone National Park (e.g. Glick *et al.*, 1991), an intact ecosystem is a system that is extremely highly integrated and that is in, or tends towards, some kind of (mostly unspecified) equilibrium, either by being so naturally or by the actions of active management. Even one of the most outspoken critics of the policy of ecosystem management and 'natural regulation' in Yellowstone, Alston Chase – although he first tries to demolish both the ecosystem concept and that of ecological equilibrium as non-scientific concepts without meaning and usefulness (Chase, 1987, especially pp. 312 ff.) – ends up in his book saying:

The eviction of the Indians, elimination of predators, introduction of exotic species of plants and animals, and a century of fire control have thrown even the 'wildest' parks into ecological disequilibrium.

And once an ecosystem has been truncated and thrown out of balance, it no longer has the capacity to cure itself. Like a seriously ill person whose vital organs are no longer functioning, these places, if left alone, will die. *(Chase, 1987, p. 382)*

Other critics, like Charles Kay – in accordance with many theo-reticians of ecology – consider ecological equilibrium to be a myth and consider changes to be 'the only ecosystem constants' (Kay, 1995, p. 107). They nevertheless urge for the reproduction of an ecological system in a particular constant state, which they see as the state of the 'original' ecosystem of Yellowstone National Park, including, however, the moulding activities of Native Americans. Kay accuses the Yellowstone managers of clinging to 'Garden-of-Eden assumptions' about ecosystems, such as the natural equilibrium of ecosystems. How-ever, it seems that even for him, 'normally', i.e. including the Native Americans, the ecosystem was in equilibrium and only in this state is it an 'intact ecosystem'.

A closer look at the documents of the NPS and Yellowstone's managers shows that there has been a changing emphasis on this issue. In the late 1960s and early 1970s, assumptions of ecosystems as equilibrium systems penetrated the writings (e.g. National Park Service, 1968), but from the late 1970s onwards (National Park Service, 1978), all explicit references to this idea vanish. There may be at least two reasons for this. First, mainstream ecological theory underwent considerable change, shift-ing slowly from equilibrium to non-equilibrium theories (Pickett and Ostfeld, 1995). Second, as for Yellowstone, the situation of the park's northern range developed in a way that made a shift in theoretical assump-tions at least convenient. Prior to 1969 the elk herd of the area had been controlled by trapping and shooting in the park, but after 1969 it was subjected to no further human control within the park. As mentioned before, Cole and Houston, research biologists at Yellowstone, introduced the idea of 'natural regulation' into the park's management policies. This idea was based at first on a notion of ecosystems as equilibrium sys-tems (Cole, 1968, 1969). However, the equilibrium numbers predicted for the elk populations were always surmounted by the actual number of elk. Thus the newer publications of the Yellowstone biologists may have, for this reason, departed from their earlier equilibrium perspective. In 1982 Houston referred to *multiple stable states* of populations instead of a single equilibrium state (Houston, 1982). Two books published by Yellowstone's biologists about the northern range during the late 1980s and 1990s – *Wildlife in Transition* (Despain *et al.*, 1986) and *Yellow-stone's Northern Range: Complexity and Change in a Wildland Ecosystem* (Yellowstone National Park, 1997) – even in their titles display this new emphasis on non-equilibrium systems. In *Wildlife in Transition*, Despain and colleagues (1986) commented:

A second dilemma involves the very character of a natural ecosystem. It is, most of all, a changing ecosystem. *(Despain et al., 1986, p. 8)*

and:

The term 'natural balance' is often a misnomer for 'natural variation.' A natural system as large and complex as the northern range [. . .] is hardly going to be the same year after year, much less decade after decade. *(Despain et al., 1986, pp. 15f.)*

In conclusion, the great overall balance or equilibrium of ecosystems ceased to be of importance for the Yellowstone biologists.

One big problem with the whole debate about the equilibrium or non-equilibrium status of ecosystems is that decisions about this issue depend on exactly what is considered to be in equilibrium and what exactly is meant by 'equilibrium'. For example, in most systems one variable may be constant over time (which is one of the most common meanings of equilibrium), despite some change going on in the system, while others might not be constant. Thus primary productivity or the ratio between primary productivity, consumption, and decomposition might be constant, while the species involved change completely – and vice versa. It seems that some of these differences make up part of the disagreement on the question of the intactness of Yellowstone's northern range.

In effect, the NPS and the managers of Yellowstone do not, as discussed above, demand an integration level as high as their critics or the environmental groups who fight for Yellowstone. All of the parties involved desire some constancy in the systems, but they differ in the degree of internal relationships they consider as necessary for this constancy, and in the variables they select and require to remain constant.

Which elements must an ecosystem have?
A second question differentiating different definitions of ecosystems refers to which elements are (minimally) considered to be present if we are to speak about an (intact) ecosystem.

One criticism of the concept of ecosystem management in Yellowstone (and other national parks) is that its application might not care about the preservation of specific species and only focus on processes instead of particular objects, e.g. species. Thus Wagner *et al.* (1995) argued about the policies of 'natural-regulation management': 'If management is focused on processes rather than entities, there need be no concern for significant changes in the states of the systems, including loss of species from our park ecosystems' (Wagner *et al.*, 1995, p. 150). Similar concerns were expressed by Chase (1987, p. 41). The definition of an ecosystem that is alluded to here is a definition that is common within ecosystem ecology, especially in circles of systems theoreticians. It considers ecosystems to be merely defined by their processes, i.e. flows of matter and energy. The different species are subsumed into functionally defined compartments and – as particular species – become unimportant and almost become invisible. The question is whether this is in fact the view of ecosystem management that the NPS and managers of Yellowstone park embrace.

The statements of these agencies – especially the NPS – leave some room for interpretation in the sense of this proposition. In particular, the 1978 management guidelines of the NPS focused to a high degree on the preservation of processes and 'wholes' as contrasted to specific species: 'The concept of perpetuation of a total natural environment or ecosystem, as compared with the protection of individual features or species, is a distinguishing feature of the Services' management of natural lands' (National Park Service, 1978, p. IV–1).

However, the perspective of the specific species was never lost and was even reaffirmed in the 1988 management guidelines, which de-emphasised the role of the 'whole' in favour of the parts:

Natural resources will be managed with concern for fundamental ecological processes *as well* as for individual species and features. Managers and resource specialists will not attempt *solely* to preserve individual species (except threatened and endangered species) or individual natural processes; rather, they will try to maintain all the components and processes of naturally evolving park ecosystems, including the natural abundance, diversity, and ecological integrity of the plants and animals. *(National Park Service, 1988, p. 4.1; my emphasis)*

The statements and writings of Yellowstone's research biologists and managers are in accordance with this. Despain *et al.* (1986, p. 17), in their book about the northern range, wrote:

One such concern is that if managers do not keep control of an ecosystem, one species of animal may outcompete others and cause the elimination of other animals or even plants. [. . .] Considering that the elk, bison, deer, sheep, and pronghorn managed to coexist here for many years before European man arrived, this concern seems unjustified. Indeed considering the differences in the habits of those animals, complete elimination of any species is unlikely. Each species has its own ecological niche to which it has become adapted. The size of each niche may vary over the years – changes in climate could significantly alter a niche – but none of them is likely to disappear completely.

The concern of the management of the Yellowstone ecosystem is (and was) not only preservation of abstract processes like primary production, the nitrogen cycle, or fire, but preservation of *all the parts*. This is the interpretation the managers of Yellowstone National Park follow regarding the law to preserve the park 'unimpaired for future generations' (John Varley, personal communication, 30 July 1998). Further evidence of this interpretation is given by the emphasis on the nativeness of the species in park management. This may be illustrated by the fact that the NPS learned with great relief that the pathogen which causes brucellosis in bison is a non-native species and thus does not need to be defended against the angry farmers of Montana who fear that the disease may be transferred to their cattle by bison leaving the park (John Varley, personal communication, 30 July 1998). A view of the park ecosystem on a more abstract, process-related level would indicate no difference between native and exotic species as long as they performed the same 'role' within the ecosystem.

Nevertheless, even if there was agreement that ecosystem management in Yellowstone focuses (also) on the particular (native) species and not just on processes, there would still be considerable differences in the perception of the ecosystem by critics and supporters of current management policies in Yellowstone.

For the managers and scientists of Yellowstone National Park, the intactness of the ecosystem is retained if all the species that were in place when the park was founded in 1872 are still there (in viable populations). In contrast to focusing on the number and identity of the species that make up the ecosystem, however, many critics of the park perceive the ecosystem and the indicators of its intactness as being on a much finer scale of resolution. For them, not only all the species have to remain as they were in pre-Columbian times, but also their specific locations and the relative abundances of their occurrences: 'Preserving the species of an ecosystem, and *in the approximate densities of what contemporary ecology considers to be reasonably intact ecosystems*, is tantamount to process management. It is also conceptually and operationally more workable for park management' (Wagner *et al.*, 1995, p. 152; my emphasis).

The aim of this kind of argument is to fix a particular state of an ecosystem in terms of clear numbers of particular species, as was done, until some time ago, in the management of Kruger National Park in

South Africa. This is manifested in the arguments about the issue of overgrazing of the northern range. Changes in the abundances and vitality of the aspen and willows of the northern range were seen as evidence of the degradation of the range and the loss of the intactness of the ecosystem. The reference state was, again, the historical state of the park as it was in 1872.

What constitutes the ecosystem of Yellowstone's northern range?
Both the managers of Yellowstone National Park and their critics remained too vague in the specification of the ecosystem concept to allow an assessment of the intactness of the Yellowstone ecosystem or that of Yellowstone's northern range on the basis of empirical data. Wagner *et al.* (1995) demanded the precise formulation of goals for the management of national parks and the variables that express what constitutes a healthy ecosystem, which they equate with the pre-Columbian ecosystem. The answer that the authors provided themselves is, however, not very encouraging as a guide on the way to a more inter-subjective definition of the properties of an 'intact' ecosystem:

Judging when the desirable state has been reached is purely a judgement call and can only be made by persons with intimate knowledge of ecosystem structure in the areas involved. The knowledge will come from long-standing personal research on the vegetation, which will include understanding the sensitivity of different plant species to grazing or browsing, or familiarity with the findings of others who have developed that understanding. It will come from exhaustive review of historical sources and photographic archives. And it will come from familiarity with vegetation in relatively undisturbed areas. Given this knowledge and that of current ecological theory and literature, it should be possible for park management to achieve and preserve intact ecosystems that bear some semblance of presettlement conditions. *(Wagner et al., 1995, p. 174)*

The answer given by the managers of Yellowstone National Park is simpler and clearer, but still not of the precision that would be necessary. Their goal, as I understand it, is to preserve all species and processes within the park with a minimum of human intervention, without trying to manage for particular numbers, ratios, and spatial relations. This is a very concrete definition of the Yellowstone ecosystem: as a spatially bounded volume with all the elements and processes it contains, without assuming that these are in any clear 'equilibrium' with each other. Given the fact that many processes and especially home ranges of the plant and animal populations within the park extend past the park boundaries, the Greater Yellowstone Ecosystem is that area which, with some probability, allows the long-term persistence of all species in the Yellowstone ecosystem without human intervention *in the park*. It is, as such, the real management unit of the park. It is not only large enough to allow seasonal migrations of animals, but also to allow for large natural disturbances such as the fires of 1988, leaving enough potential for recolonisation of disturbed areas.

There are, however, some unproven assumptions and problems involved in the current management concept of Yellowstone National Park. It is an assumption, based on a lot of ecological and historical knowledge, but not in any way certain, that 'natural' changes occurring in the area will nevertheless allow the persistence of all species within Yellowstone. Allowing natural dynamics, as is done to a high degree at present, might in fact lead to the loss of one or the other species. The park management acknowledges this possibility and would allow for this, if the extinction was based on natural processes (John Varley, personal communication, 26 August 1999). This raises several questions: what, again, is 'natural'? And where is the point of intervention for Yellowstone? How many species may go extinct without endangering the 'intactness' of the ecosystem if this is defined in terms of the particular species and the processes in which they are involved? At what point would the 'experimental management' of the northern range (Houston, 1981) be considered to have failed? Is climate change a natural or an artificial process? How should the NPS react to it in either case?

The case study shows that it is far from trivial to agree on what constitutes an intact (or functioning) ecosystem. It also demonstrates how complex scientific, philosophical, and political arguments are intertwined when it comes to assessing ecosystems. Empirical data play an important, if not the predominant, role in such debates and most arguments are made with the (assumed) authority of science. But in fact it is not scientific data alone that guide the arguments about the intactness of ecosystems – even though this may often not be obvious and/or conscious. Ideas of nature, assumptions about how nature 'works', about the role of humans within nature, about desirable historical states of an area and how far they are reference states for 'intact' ecosystems, all shape our arguments about what intact, healthy and functioning ecosystems are. Many of these issues rely heavily on values (an issue I will come back to in Chapter 5). There are also a number of contested assumptions from within ecological theory involved. But to a considerable degree it is the vagueness of the ecosystem concept – or a lack of more sufficiently specific definitions of it – which, on the scientific side, shape judgements about ecosystem functioning. So to really assess ecosystem functioning, we need to clarify our ideas about ecosystems and communicate them in a way as unambiguous and conceptually sound as possible.

4.3.2 A tool for clarifying and visualising different ecosystem definitions: the SIC scheme

If there are so many different definitions of 'ecosystem', how can we make sure we use clear language about them, and how can we classify them? While an in-depth review of the different ecosystem concepts is beyond the scope of this book and has been done in other places (Jax, 2002a, 2006), it is necessary to provide at least an idea about how we can clearly communicate about what we mean when we use the term 'ecosystem'. For this purpose, I will use the so-called SIC scheme, as developed by Jax *et al.* (1998) (see also Jax and Rozzi, 2004). I will only briefly sketch the approach – as far as it is useful and necessary in the context of this book (see Jax, 2002a for an extended treatment of this approach).

The scheme is called the 'SIC scheme' after three of the criteria on which it builds to define ecosystems or other ecological units: **S**elected phenomena, **I**nternal relations, and **C**omponent resolution (see below). It was developed in response to the question of how we can assess whether an ecological unit (such as a population, a community, or an ecosystem) is still the same after a period of time, if it has become another unit (changed its 'identity'), or if it has been destroyed. As everything in nature changes almost all the time (such as individual organisms being born and others dying), this is not a trivial question, but one which aims to distinguish essential characteristics of a system from 'accidental' characteristics. The differences in definitions of the ecosystem of Yellowstone's northern range, as discussed above, have provided some good examples of such distinctions. The SIC scheme is a conceptual model for expressing the meaning of 'ecosystem' (or any other ecological unit) in an unambiguous manner. It sorts out four basic questions that we have to answer when speaking more specifically about ecological units.

The basic parts of any ecological unit are its *components* (or elements), of which it consists, and (in most cases) the *internal relations* (processes/interactions) between the components. The components and

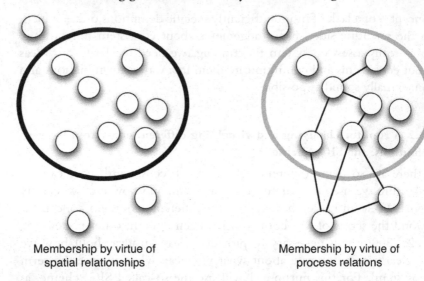

Membership by virtue of   Membership by virtue of
spatial relationships   process relations

Fig. 4.8. The two boundary types of ecological units: topographical boundaries
(left) and process-based boundaries (right). See text.

internal relations are what I call the *phenomena* which constitute the unit.
Besides this, every unit has a boundary in space.

To provide an unambiguous definition and specification of any ecological unit, statements are needed about:

(1) whether the unit is bounded topographically or process-related;
(2) which kind of relationships among the components are minimally
    required;
(3) which phenomena (i.e. components and internal relations) are
    selected for the definition of the unit; and
(4) what is the degree of resolution of the unit's components.

Boundary criterion: topographical or process-based boundaries
While the other criteria may be thought of as gradients, the two kinds of
boundaries which ecological units can have form a clear dichotomy. The
boundaries of a unit can be topographical or process-based (Fig. 4.8).
The same fact can also be expressed in that elements of a unit may, in the
first case, be perceived of as parts of that unit by virtue of being spatially
related (i.e. their presence is in the same locality), and in the second case
by virtue of their linkage through interactions. An example of a community defined by spatially related elements is all of the organisms occurring
in a particular lake. In contrast, an example of a community defined by

interactively related elements is organisms in a lake engaged in a particular kind of interaction, such as trophic relations. Although interaction is sometimes inferred from proximity, such correlations can be spurious. A topographical boundary does not need to have significance in terms of process-based boundaries and vice versa. There should be a clear statement as to what is essential for elements to be a part of an ecological unit.

Topographical delimitations of ecological units sometimes deal with boundaries drawn on the basis of purely practical considerations (sampling plots, administrative boundaries). In all other cases they are based on one or more *homogeneity criteria*, with the aid of which one area is made distinguishable from an adjacent one. The boundary is a break in this homogeneity. Topographical boundaries are 'tangible'. In contrast, process-based boundaries are not tangible; like all empirical boundaries they have a spatial dimension, but they are not 'fixed' as clearly in space as topographical boundaries. For process-related boundaries, the boundary-'surface' is instead given by a gradient of interaction strength. This means that with respect to a particular definition, the interactions between the elements of a unit with each other are stronger than those with other elements, which belong to their 'exterior' (Platt, 1969; Allen and Starr, 1982; Ahl and Allen, 1996). An example may be an orchestra, where the interactions of the musicians with each other are stronger than their interactions with the audience or with other orchestras.

In both cases, however – also for units delimited on the basis of processes – the boundaries are determined by the observer. The observer chooses the homogeneity criteria for the topographically delimited unit or draws the boundaries according to purely practical considerations. In the case of a unit delimited on the basis of processes, he or she has to fix the relevant process/interaction variables and the threshold values which determine at what point a gradient in the intensity of interactions is sufficiently steep to form a boundary. To put it another way: how strong must an interaction be minimally to consider a potential element as being connected with the unit? Grizzly bears regularly eat fish from Lake Yellowstone in Yellowstone National Park. Are they thus part of a lake species assemblage defined by food relations? The answer to this question is in no way obvious.

The other criteria introduced below can be applied in the same manner to either topographically bounded units or units bounded on the basis of processes. Due to their gradual character, these criteria can be depicted as axes, which provide information about the characteristics of the ecological unit to be defined.

The degree of internal relationships (I-axis)

The first axis on which a particular definition of an ecological unit can be arranged is that of the degree of expected internal relationships between the elements. By this, I mean the degree of relations between an assemblage of elements (e.g. species), which is considered as necessary to speak of an ecological unit in the defined sense. As with the other axes, this is not a strictly quantitative gradient that can be expressed by one simple variable and measurement value. In fact, all the axes are complex in themselves (see Jax, 2002a).

Ecological units are sometimes defined simply by the common occurrence of several organisms in a chunk of space; they may also be defined as requiring specific patterns, i.e. characteristic, recurring combinations of particular elements or types of elements (e.g. species lists), or they may be defined by process relationships, in which the interactions become the focus of the definition (e.g. trophic relations, competition, etc.). The first approach is frequently applied if ecological units are used for classificatory purposes, such as the mapping of plant or animal communities, or the classification of ecosystems (e.g. Klijn and Udo de Haes, 1994; Bailey, 1996). Investigations aimed at the prognosis of the dynamics of ecological units generally embrace process relationships as a decisive criterion for definition (e.g. Odum, 1983).

The viewpoints are not mutually exclusive. On the contrary, patterns and interactions are mostly interrelated. Note, however, that while most definitions of communities based on recurrent species occurrence imply at least some interactions, a process-focused approach does not necessarily imply specific species combinations (see also O'Neill *et al.*, 1986).

Different types and degrees of internal relationships among components are thus required in different definitions of ecological units (Fig. 4.9). One extreme viewpoint of the definition of a unit, for example, a population, requires no interactions between the elements in order to be called a population, while the requirements in respect of the intensity and specificity of interactions grow along the axis, and are needed to call an assemblage of organisms of the same species a population.

In the case of species assemblages, a definition that requires no interactions between the elements of the community leads to a purely statistical or pattern-based definition of a unit. In this case, an assemblage of elements is called an ecological unit whenever these elements occur in a particular combination. A classical example is given by Petersen (1913) in communities of marine benthic organisms – which, in more sophisticated forms, is still used (e.g. Warzocha, 1995; Zenetos, 1996). Moving

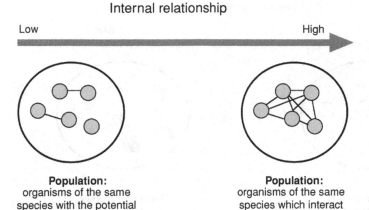

Internal relationship

Low　　　　　　　　　　　　　　　　High

Population:
organisms of the same
species with the potential
to interact

Population:
organisms of the same
species which interact

Fig. 4.9. The axis of internal relations (I-axis); see text. Figure reprinted from Jax *et al.*, 1998, with permission from Oikos.

along the gradient, away from the extreme of no interactions required towards a more interaction-oriented definition, the range of possibilities is very broad. An assemblage of species is frequently considered as a community if there are only (or at least) some (any) unspecified interactions between the organisms of the different species (e.g. Abele *et al.*, 1984). 'Intermediate' definitions would require, for example, specific connections between all elements of the unit, although they may not necessarily lead to any constant 'equilibrium' pattern. The most stringent requirement for an internal relationship comprises equilibrium, self-regulation, or the autonomy of each particular assemblage (e.g. the 'superorganisms' of Clements (1916) or Lovelock (1979)). The emphasis of the gradient expressed in this axis is not on the number of elements that are connected by process relations (this is part of the axis of 'selected phenomena'), but more on connectivity, intensity, and, especially, specificity of the inter-relationships.

Please note that a definition of a community or an ecosystem that is located near the lower end of the I-axis does not mean that stronger or more integrated relationships cannot or must not *exist*. They are just not important for that specific definition of an ecological unit.

Selected phenomena (S-axis)
The second axis, termed the selected phenomena axis (Fig. 4.10), relates to what kinds of elements and how many elements and interactions have

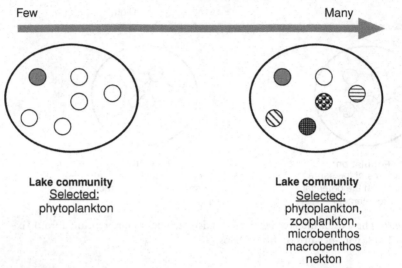

Fig. 4.10. The axis of selected phenomena (S-axis); see text. Figure reprinted from Jax *et al.*, 1998, with permission from Oikos.

to be present in a unit to satisfy its definition. Do all organisms in a particular location belong to the species assemblage (or ecosystem), or is it sufficient for only some to be there? Thus, a species assemblage does not need to comprise all the organisms in a lake, but might only consist of all fish or all protozoa of a lake, or might be restricted to a particular size class, for example the macrozoobenthos assemblage of a lake, as opposed to the microzoobenthos assemblage. In such cases, different assemblages or ecosystems defined by a low number of selected elements (a subset of a theoretically higher number of potential elements) may overlap in one volume of space (see also Ahl and Allen, 1996, pp. 25f.). The phenomena can be selected elements, such as taxonomic or size groups, and/or processes – for example: food relations, fluxes of matter, etc. For example, an ecosystem could be specified by 'all' functional relations occurring between the elements, or only by some of them, such as particular pathways of gas exchange, or some hydrologically mediated relations.

Like the internal relationship axis, the selected phenomena axis is a complex one. The distance of any point from the origin relates both to the number of phenomena selected ('few' to 'all') and to a qualitative component, as different kinds of phenomena can be selected. For

Component resolution

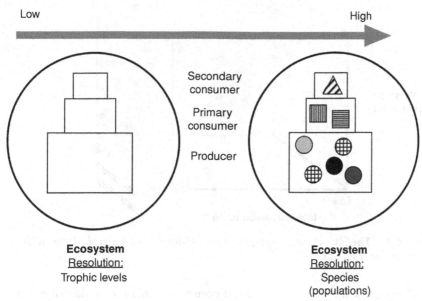

Fig. 4.11. The axis of component resolution (C-axis); see text. Figure reprinted from Jax *et al.*, 1998, with permission from Oikos.

example, if only one interaction variable is required in the definition, it might be either the phosphorus cycle, the nitrogen cycle, or some other process. A particular phenomenon can have different expressions or values. Thus, if only higher plants are included in the specification of a species assemblage, different types of assemblages may be described, such as a beech–oak forest assemblage or a spruce–fir forest assemblage. Likewise, specifying a lake ecosystem based on the inclusion of nutrient and oxygen dynamics might distinguish oligotrophic and eutrophic lake ecosystems.

Component resolution (C-axis)
The degree of component resolution (or its opposite, degree of aggregation) indicates the generality under which an element is perceived. It is arranged on the third axis (Fig. 4.11). At a low degree of resolution, the organisms of a lake or forest might be aggregated to trophic levels or size classes. One might, for example, consider nitrogen flow between producers, consumers, and decomposers. In the extreme, even variables such as 'biomass' might be sufficient to characterise the ecological unit. At the

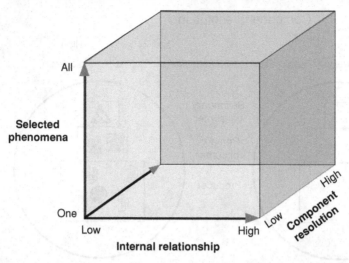

Fig. 4.12. The SIC scheme; see text. Figure modified from Jax *et al.*, 1998, with permission from Oikos.

other end of the scale, a fine–tuned component resolution describes particular species (or even finer subdivisions) as the elements of the system. Here, to use the above example again, nitrogen flow would be measured between particular species and not just between aggregates of species.

A three-dimensional graphical model: the SIC scheme
The axes described above can be assembled into a three-dimensional graphical model that lays out an abstract volume in which any ecological unit can be localised (Fig. 4.12). As noted above, all the axes themselves are complex. Thus, theoretically there may be a further division or even a hierarchy of axes (Jax, 2002a), so investigators may use slightly different or more detailed criteria, which nevertheless can also be subsumed into the broad classes given here by the three axes. These axes, together with the boundary criterion (which is not displayed, but has to be stated explicitly), state the things that ecologists must specify minimally to give a definition of the specific ecological unit that can be communicated inter-subjectively, and which, in particular, ensures the connection of empirical data and theoretical ideas in this field.

The four criteria described (boundary criterion, internal relations, selected phenomena, and component resolution) are the result of an analysis and systematisation of existing definitions of ecological units (Jax, 2002a, 2006). They reflect – grounded by the reference to concrete

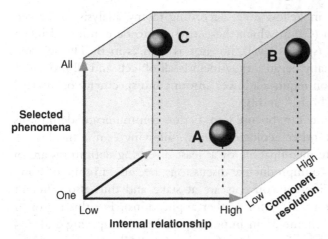

Fig. 4.13. The SIC scheme as applied to the northern range controversy. See text.

ecological research – the fundamental general characteristics of ecological units as systems.

The specification of the elements forming the systems and of all interactions relevant within the systems is provided by the axis of selected phenomena (S-axis). The interactions themselves are determined in more detail by the axis of internal relations (I-axis). Finally, the resolution of the elements is specified in more detail by the axis of component resolution (C-axis). The model provides information about four decisive system characteristics in an ecological context. The name, 'SIC model', uses the abbreviations of the three axes as a shortcut.

Application of the scheme to the Yellowstone controversy
Figure 4.13 uses the SIC scheme to illustrate the different definitions discussed for the Yellowstone ecosystem in Section 4.3.1. In fact, I have already discussed the differences between the definitions of the ecosystem in the discourse about Yellowstone's northern range (Section 4.3.1) along the lines of this scheme.

The critics see the NPS policy in Yellowstone as aimed at a system (A) with a low component resolution (single species being of minor importance); an intermediate amount of variables (mainly processes); and high internal relationships. This is, however, for them, not a description of an intact ecosystem. The definition they favour (B) sees an intact ecosystem with a very high resolution of components; a rather high amount of selected phenomena (all objects which were present previously, but specific processes are less relevant); and high internal relationships. In fact, the NPS strategy in Yellowstone, according to my analysis, embraced another definition (C), in which the ecosystem to be considered intact is highly resolved (species level), but not as high as in B. The selected phenomena are many because they include all objects and all processes, while the definition requires a lower amount and specificity of internal relationships than both A and B.

The SIC scheme can be used to better communicate definitions of ecosystems (or other ecological units), allowing – not the least – visualisation where incompatible or at least diverging definitions are in danger of leading to unproductive discussions, because, in spite of using the same word, different concepts are at stake and thus also different objects (see Jax *et al.*, 1998 for further applications). For this reason, I will use the SIC scheme as an important tool for comparing and discussing different ideas of ecosystem functioning in Chapter 7.

4.4 Conclusions from this chapter

4.4.1 General conclusions

The main message of this chapter is that when talking about ecosystem functions or ecosystem functioning, we need more clarity with respect to what we mean by 'function(ing)', as well as with respect to what is our target system, i.e. what we mean by 'ecosystem'.

First, we should at least be explicit as to whether we refer to ecosystem *processes*, the overall performance of an ecosystem (*functioning* in the strict sense); to the role (purpose) of an ecosystem or of parts of it (*function* in the strict sense); or the *services* derived from an ecosystem, as those processes that are demanded to provide human benefits. I will reserve 'ecosystem functioning' in the following text to exclusively refer to the overall performance (or operation) of an ecosystem. The question of how to make this notion of the 'overall performance of ecosystems' more concrete – that is, how to operationalise (or measure) ecosystem functioning – will be the subject of the remainder of this book.

As I have said, 'functioning' alludes to some normative state ('proper' functioning) of a system or even to some goal or purpose of the system. This becomes even stronger when we reverse the order of words, speaking about a *functioning ecosystem* being the manifestation of ecosystem functioning. We should be aware that, for ecosystems, there is no inherent purpose or goal and that any references to such goals can, at best, be metaphorical mirroring goals that an observer sees or desires in a system (notwithstanding that the specific states of ecological systems are not purely a matter of perspective).

The second necessity when assessing ecosystem functioning is to be explicit in what we mean by 'ecosystem'. An ecosystem is not given by nature as a fixed entity that we just have to 'discover' or 'identify'. From the same physical segment of nature, different objects can be derived as 'ecosystems'. There is not a single 'correct' definition of an ecosystem, but there is instead a plethora of different possibilities to define and delimit an ecosystem in a scientifically sound manner. In Section 4.3.2 I introduced a framework to clarify the different definitions of an ecosystem (the SIC approach). This approach (or similar ones) can help to formulate ecosystem concepts in a precise and unambiguous manner. It should be evident that for different objects (and different definitions which 'extract' different physical objects from the same whole of nature), different answers with respect to the question of whether they are 'intact' or 'functioning' may result.

The definition of a functioning ecosystem, starting from the question of which of the many possible definitions is selected and appropriate, involves *by necessity* choices, and even value dimensions. These dimensions are often not made explicit, nor are they always conscious. As we have seen from the Yellowstone case study, they include specific world views, specific assumptions about what and how 'nature' is, the specific purposes of study, but also assumptions from ecological theory. It is thus not a purely scientific issue, but involves societal choices.

We might now conclude that if ecosystem functioning implies normative dimensions it is not a useful scientific concept at all, and that we should in consequence abandon it or should use it only in the broadest metaphorical sense. Also, if there are so many possibilities for defining ecosystems and their (proper) functioning, and if the choices to be made cannot be made on scientific grounds alone, does it mean that the ecosystem and ecosystem functioning are only socially constructed? Is ecosystem functioning a scientifically empty concept and its contents completely subjective and arbitrary? How do societal choices and scientific facts relate to each other, if they are both parts of determining what ecosystem functioning means?

The issue that is raised by these questions has two aspects: a theoretical and a practical one. The theoretical question is as to whether a subjective element renders a concept arbitrary, 'unscientific', and purely 'constructed'. If we can reject this proposition (as I will do below), we still have to clarify the role that values and choices play in the context of conceptualising and using the concept of ecosystem functioning. These are the issues that I will deal with in Chapter 5. If we find that the concept can be formulated in a scientifically and philosophically sound manner, and that it can be useful for ecology and conservation biology, the question arises of how to connect societal choices and scientific facts in practice. That is: how do we select an adequate concept – adequate in terms of corresponding to physical reality as well as adequate for using it for specific scientific and societal purposes – and how do we operationalise it? This will be dealt with in Chapters 6 and 7.

Before proceeding further, however, a few more words must be said about the domain of terms and concepts with which we deal later.

4.4.2 Ecosystem functioning as a conceptual cluster: related terms and concepts

Even if we perceive ecosystem functioning in the more narrow meaning, there is in fact a large number of related terms and qualifiers in the literature which express the same idea: namely that an ecosystem as a whole operates in a specific manner, which, within some limits, can be described as a reference condition (state, trajectory, or dynamics). As I said before, this book deals not just with the *term* 'ecosystem functioning' and its various meanings, but with a particular idea, the *concept* behind most uses of the word. This is expressed by phrases such as 'good functioning', 'proper functioning', 'correct functioning' of ecosystems; by describing ecosystems as 'fully functioning' or 'malfunctioning'; by talking about different 'degrees of functioning'. The same idea is at stake when characterising an ecosystem as (fully or not) functional, healthy, recovered, damaged, degraded, restored, cultivated, self-sustained/self-sustaining, or when discussing their 'integrity' and 'resilience'. All together these terms form what I have called a *conceptual cluster* (Jax, 2006; Jax and Schwarz, 2010). Other terms and phrases that belong to this cluster are, for example, 'total ecosystem function', 'ecosystem reliability', 'ecosystem failure', 'ecosystem viability', 'ecosystem thresholds' and/or 'tipping points', as well as 'critical limits of biophysical integrity'. In detail, these different expressions imply quite different ideas about the 'nature' of ecosystems and the ways to understand or assess their functioning; but they converge on a kind of normative state as to the performance of the whole system. It is one task of this book to view these related concepts together and analyse their common problems and potentials. From that basis, we can better explore their differences and, in particular, scrutinise the possibilities of *applying* them in practice – in a philosophically sound manner.

5 · Ecosystem functioning
Science meets society

Dealing with 'ecosystem functioning' is not just a matter of science; its meaning and assessment are also dependent on societal choices. This is one of the basic theses of this book. In this chapter, I will take a closer look at what it can mean that ecosystem functioning is (also) a matter of societal choice. For this purpose, I will first (Section 5.1) investigate a possible conclusion that one might draw from the previous chapter – namely that ecosystems are mere (social) constructs – and the implications that would be connected with such a conclusion. By means of a case study about so-called alternative stable states of ecosystems, I will demonstrate that ecosystems and their functioning are at least not *mere* constructs, but are situated between scientific descriptions and societal choices. As societal choices are always connected with values, I will then discuss how (and which) kinds of values enter into the process of defining and assessing ecosystem functioning (Section 5.2). In another case study (Section 5.3) I will describe how societal choices are part of ecosystem management strategies, analysing especially the so-called Ecosystem Approach (EA) of the Convention on Biological Diversity (CBD). Section 5.4 will then present some general conclusions on the roles of science and society in assessing ecosystem functioning.

5.1 Between constructivism and scientific realism: determining the limits of ecosystem functioning

5.1.1 Are ecosystems mere constructs?

The statement that the term 'ecosystem' can be meaningfully defined in many different ways may be interpreted as a kind of constructivistic perspective towards nature, and as a relativistic position. My colleague Clive Jones, for example, during our common work about the self-identity of ecological units (Jax *et al.*, 1998) once quipped that we were doing 'relative ecology'. So, are ecosystems just relative, are they constructs, mere figments of the mind? What, then, could 'ecosystem functioning' mean at all? Does it mean that 'anything goes' with respect to resource and ecosystem management – that ecosystem functioning is just a matter of personal opinion and interpersonal agreement? The answer to these questions is crucial to our understanding and use of the concept of 'ecosystem functioning'. The question of whether ecosystems are constructs cannot be answered simply by either 'yes' or 'no' (in fact, both are correct), but requires a closer look at a sometimes very heated controversy, during which many strawmen have been erected and burnt.

The position of constructivism, or more specifically the claim that particular things, such as gender, emotions, danger, quarks, wilderness, and even nature as a whole are socially constructed (see Hacking, 1999, p. 1 for a list of things that are said to be socially constructed) has aroused many controversial discussions, both for philosophical and political reasons. The issue behind these controversies is of considerable importance for conceptualising and operationalising ecosystem functioning. In particular, it can cause major misunderstandings if not dealt with in a differentiated manner. In order to stay focused, I will confine my treatment of this issue largely to discussions that refer to the realm of ecology and conservation biology.

So let us look at what it may mean to say that nature or – in the specific case of this book, ecosystems and ecosystem functioning – are socially constructed. I will start by exposing the issues behind this statement by referring to the well-established (and still moot) discussion on the social construction of nature, and then proceed to the concepts of the ecosystem and ecosystem functioning.

The extreme positions towards what nature 'really' is and how we can talk about it, are those of a *naive realism* on the one hand and a *strong constructivism* on the other. Proponents of a naive realism believe that we are able to directly perceive nature through our senses as it really is and that this reality is independent of human perspectives, interests, and values. In contrast to this, strong constructivism negates the existence of facts and nature as such and postulates that nature is a purely cultural phenomenon, depending on specific cultural perspectives on the world. Most philosophers of science reject naive realism, claiming that our perceptions of nature (or reality in general) are always filtered and influenced by our previous cultural backgrounds and experiences (previous knowledge, theoretical assumptions, values, etc.) and by our interests. But there is also strong criticism of the constructivist position from philosophers, but even more from the camps of conservation biologists and conservationists. Part of this critique is political: if nature is just a human construct, there is, they state, no reason to protect any specific state of the world. More than that, the discussion about our perceptions and modes of construction of nature also distract us from solving the ecological crisis or even questioning whether there is a crisis at all (Soulé and Lease, 1995; Kidner, 2000; Crist, 2004). Another part of the critique is philosophical: how is it possible to negate the very existence of nature as existing independently of us and our ideas, interests, and attitudes?

In his book, *Pandora's Hope*, the French sociologist Bruno Latour, being one of the most cited 'constructivists', recounts a meeting with another scientist (a psychologist) who had read his writings. 'Do you believe in reality?' he was asked. Latour, somewhat surprised, amused, and embarrassed at the same time, replied: '"But of course!" I laughed. "What a question! Is reality something we have to believe in?"' (Latour, 1999, p. 1). This little episode displays quite well the kinds of misunderstandings that exist with respect to what it means that nature is socially constructed. As several authors (Proctor, 1998; Hacking, 1999; Demeritt, 2002) have demonstrated, the statement that something (nature, ecosystems, and other things) is socially constructed − or the accusation that someone considers something as 'just constructed' − is prone to misinterpretation, because different things are meant by 'socially constructed'.

First, 'constructed' may refer to different things. Ian Hacking (1999), in the title of his book asks: *The Social Construction of What?*. The 'what' may either be an object, a concept, a classification, or an idea ('out in the public'). He and others agree, however, that the most important distinction here is between the construction of a concept and that of an object. This also alludes to a distinction between epistemology and ontology. Do we mean that a *concept* of something is/has been socially constructed or is the *object* as such constructed and does not exist without humans? Do we mean, as two ways to understand the latter, that we have constructed the object physically, by physical human actions, or by our mental acts? There are considerable differences between different groups of 'constructivists' with respect to these questions (Hacking, 1999; Demeritt, 2002). While a few constructivists would really argue that language and social processes are the only 'real' things, preceding any physical reality, most constructivists, as the quote from Latour demonstrates, do not deny the existence of some kind of reality existing independently of us, even though they strongly question how much we can really know about it in terms of 'objective' knowledge.

There is also often a close relation between conceptual and physical construction of nature (and other things). Humans, of course, have always modified and thus constructed nature (here: the physical world) and at the same time imagined (constructed) what nature *is* (to them). As Demeritt (2002, p. 779) emphasises: 'Clearly many material constructions of nature will depend on the conceptual constructions that guide the ways people interact with and transform the physical environment, which in turn will influence what people conceive.'

Furthermore, there are different categories of what 'nature' means to those who argue about its construction. The debate is in fact not about the existence or non-existence of nature in the more generic and extreme meaning of nature as everything that exists, physical nature, the 'world outside' (Demeritt, 2002, p. 778; see also Rolston, 1990). It is (mostly) a strawman erected by opponents of constructivism (or 'deconstructivism', as they sometimes call it) that this kind of nature is seen as denied or is considered as being constructed by constructivist philosophies. Most of the proponents from the constructivist side argue about specific expressions of the 'nature of nature' (i.e. what it means of something to be 'natural') and our (in)ability to describe or recognise it with certainty. In fact, the discussion is one about *epistemology*, about our ability to know about nature and not so much about *ontology*, i.e. as to whether nature exists and what it 'really' is (Proctor, 1998). Or, as Phil Kitcher put it succinctly: 'So do we construct the world? In the sense often intended in fashionable discussions, we do not. There is all the difference between organizing nature in thought and speech, and making reality: as I suggested earlier, we should not confuse the possibility of constructing representations with that of constructing the world' (Kitcher, 2001, p. 51).

But even such moderate (epistemological) claims are attacked fervently and also often mixed up with an ontological perspective:

Such [constructivist] claims suggest that nature is an entity very different from that which many environmental theorists, writers, and activists have up to now believed. Rather than being viewed as a multifaceted, diverse order whose pattern and possibilities extend well beyond our abilities to understand them, nature becomes an offshoot of *social* reality which also constructs individuality. And since the social world varies according to time and place, then it follows that each of these social worlds will construct a somewhat different version of nature, and there is, therefore, no single 'nature', but rather a 'diversity of natures', constructed by our various fantasies and languages. *(Kidner, 2000, p. 340)*

But interestingly, the discussions that constructivists lead about the nature of nature (in their case from an *epistemological* point of view) are also at the heart of many conservation controversies. However, they are led here – mostly from a realist position – discussing the *ontology* of nature: what 'real' or 'true' nature is (see the discussion about the intactness of Yellowstone's ecosystems, described in Section 4.3). Although these controversies are thus not framed as being about the social construction of nature, as they are looking for *the* nature of nature, the point of departure is the same: the fact that different perspectives prevail about what exactly 'nature' is.

Neither naive realism nor extreme constructivism are really satisfying when it comes to discussing nature and ecosystem functioning. There are thus many authors who, in different ways, argue for a middle position between these extremes (e.g. Bird, 1987; Pickering, 1995; Gandy, 1996; Proctor, 1998; Kitcher, 2001). My own position also dwells within this middle ground. Although I acknowledge the issues posited by a constructivist view as meaningful and important with respect to nature and ecosystems, I prefer to express it in other words. Of course, nature (as the physical world) is 'out there' (in an ontological sense) – including ourselves in so far as we are physical beings. The way we can perceive nature, however, what we can know about it (the epistemological dimension) is clearly in our minds. It is influenced by our cultural backgrounds (in the broadest sense of the term) and by our specific interests. In other words, nature is *also* socially constructed. It is a concept that describes our relation to the world. Nature is a concept originating from human minds, but at the same time it is something 'out there', to which our concepts refer, and which the sciences, such as ecology, describe, as do lyricists and painters. This 'something out there' is largely independent of us. We cannot know nature in totality (its essence, or what it 'really' is – the 'things as such', as Kant expressed it), but nevertheless, physical nature and its parts resist any purely arbitrary definitions. All the conceptual constructs we make of it, when applied to a specific situation, have to pass a test – namely whether they fit the situation – for example, in terms of predicting the dynamics of the things we refer to in our concepts, or even manage them. We may construe a concept of a stone wall as an easily permeable boundary between two places, but as soon as we test this definition by trying to move through the wall, the resulting headache suggests otherwise. So usefulness in a practical, pragmatic sense is one test that concepts should be able to pass. As Pickering (1995, p. 22) puts it, in the course of practice nature *resists*, forcing us to accommodate the concepts we construct about it (several authors have pointed out that the word 'object' – or 'Gegenstand' in German – from its etymological roots contains the notion of resistance: 'to object', to 'stand against'). A similar idea was formulated by Elizabeth Bird, saying that we are, in doing science, *negotiating* not only about the interpretation of our observations, but also about reality. As she expressed it:

[R]eality *is* being negotiated at the same time as its theoretical construction. And both of those, the reality and the interpretation, are not merely social constructions, but at both levels negotiations with nature. Nature's role in that negotiation takes the form of actively creating something materially new and resisting or accommodating the range of metaphorical and theoretical imaginings with which it is approached. *(Bird, 1987, p. 259)*

5.1.2 Implications for ecosystems and their functioning

So far I have spoken mostly of 'nature' in general. Let's now come back to ecosystems and their functioning. In the most general (generic) meaning 'nature' and 'ecosystems' exist. Nature – taken as the totality of things 'out there' – on earth today involves a web of organisms of different species interacting to various degrees with each other and with their abiotic environment. Chunks of this web, as well as the web as a whole (the biosphere), are what we call 'ecosystems' in the most general meaning of the word. Almost all ecologists will subscribe to this statement. But with this generic meaning of 'ecosystem', what is ecosystem functioning and what is a *functioning* ecosystem? The generic meaning of 'ecosystem' is more a kind of *perspective* on nature, looking at nature as a web of organisms and their environment interacting with each other (see Box 5.1). As such, it is in fact so vague and general that only with no organisms left, would an ecosystem cease to function. It is only with more narrow definitions of 'ecosystem' that we can apply the concept of functioning in a meaningful manner. That is, we have to divide the biosphere into smaller units with more specific criteria telling us when we consider a particular segment of the biosphere to be an ecosystem, or only part of an ecosystem (see Jax, 2006 and Section 4.3). As a next step we must then go on to decide under what conditions we consider it to be a (properly) functioning ecosystem, a malfunctioning ecosystem or a destroyed ecosystem, requiring the determination of specific reference conditions. A question, then, is: can we find 'natural' subdivisions of the biosphere to classify into ecosystems of more specific kinds?

Box 5.1 *The ecosystem as an object and as a perspective*

The term 'ecosystem' is not always used to describe a narrowly defined concept, as delimiting a specific unit in space and time. In fact, it is both used as a *perspective*, or general heuristic idea, and as a concrete *object* (unit). As a perspective, in its generic definition ('an assemblage of organisms of different types together with their abiotic environment in space and time' (Jax, 2006, p. 240; see also Chapter 4)), it emphasises the interrelatedness of ecological phenomena, especially that between biotic and abiotic elements of our world. Such general, generic terms are useful and necessary for science, in spite of, or even *because* of their vagueness, because they can help to structure a field of research – in this case, ecology. That is, they provide a rough organisation of research objects and/or research topics and facilitate, understood in this way, the process of communication. In the generic manner, 'ecosystem' roughly delimits a particular class of phenomena, which may also be designated as an *observation level* or *perspective*, in contrast to the more specifically defined ecological *units* (Jax *et al.*, 1998; see also Wiegleb, 1996). The problem of the ambiguity and limited applicability of these concepts nevertheless persists, especially when the distinction between perspective and object becomes blurred. In consequence, the term 'ecosystem' (and likewise 'population' and 'community') should only be used as a very broad generic term (e.g. in the sense of the definition mentioned above). More specific uses of the term should be indicated through an explicit definition. Such more specific uses are required for theoretical generalisations and for operationalising the ecosystem concept in practice.

Multiple roles of the ecosystem concept have also been characterised by Pickett and Cadenasso (2002), who describe the uses of the ecosystem concept as *meaning, model,* and *metaphor.* The first dimension (meaning) comes close to the generic use of 'ecosystem'; the second (model) to more specific definitions; and the third (metaphor) as an informal and symbolic use of 'ecosystem', also in a management or policy context. This corresponds with the observation that the 'ecosystem' is used in recent approaches of 'ecosystem management' both as an object and a perspective, the latter here describing a kind of policy or philosophical approach, extending beyond the bounds of pure ecology towards the social sciences (Jax, 2002b).

Similar distinctions of different use categories for ecological concepts have been described by King (1993; for the ecosystem concept) and Gaston (1996; for the biodiversity concept). For an extended discussion on the role of broad and imprecise concepts in ecology, see Jax (2002a, 2006).

Up to now, there have been no uniform and generally agreed-upon definitions of ecosystems which can also really be applied in practice (e.g. for assessing ecosystem functioning). Instead there is a plurality of definitions, most of which are useful and adequate for at least some purposes (see Jax, 2006). Thus, different ecologists, faced with the task of distinguishing the ecosystems in a particular area and/or counting them (Cousins, 1987), will come up with different results, depending on his or her perspective, theoretical background, and interest. In contrast, there would be much more agreement if the same scientists were given the task of distinguishing and counting the individual organisms (e.g. birds) in the area. This is not a matter of the progress of ecology, but brings us back to the question of the constructedness of ecosystems.

To deal with the impressions our senses convey to us, we have to make distinctions, i.e. we have to classify the world into objects of certain types. Some of these classifications appear to be made more easily, such as when they correspond to our everyday experiences. It is also helpful when different approaches of classifications lead to concepts with very similar objects included. Such classifications and the concepts that derive from it are, as Allen and Hoekstra (1992, p. 26) called it, 'robust to transformation' (see also Jax, 2007). Examples of such concepts are the 'individual organism' or the species concept in biological taxonomy. But ecosystems are not 'robust' concepts in the above sense. Nor are they something like 'natural kinds', i.e. classifications of things that are beyond our specific interests and theories. The idea that nature is neatly divided into parts that we have to discover is directly related to an understanding of an ordered world. It is certainly an appealing notion that nature would provide us with inherent classifications, but it hardly works. In biology, the species concept has often been discussed as a candidate for natural kinds. One argument is that folk classifications and scientific classifications of bird species match very closely (Mayr, 1988b, p. 317). However, even for the species concept this claim is highly contested (e.g. Dupré, 1992; Rosenberg, 1985; Ruse, 1987). Even within biology there are many different species concepts (e.g. biological, genetic, evolutionary, and morphological concepts) which do not always yield the same result if applied to the same organisms. Even the concept of the (individual) organism, which seems rather clear and obvious to us at first sight, is not a clear-cut and 'natural' category (Sterelny and Griffiths, 1999, pp. 70f.). In the case of counting clonal organisms, for example, it is not evident what the individual organism is (Janzen, 1977). Is, for example, a single shoot of clonal plant such as *Phragmites*, visible above ground, the individual? Or is the individual instead the whole clone, composed of all those reed shoots which are morphologically and physiologically connected with each other through their roots? Here a decision has to be made if ramets (structural individuals) or genets (genetic individuals) (sensu Harper, 1977) are meant.

Accordingly, Kitcher, otherwise a self-declared 'moderate realist', states with respect to the notion of natural kinds: '[T]he thought that nature comes with little fenceposts announcing the boundaries of objects is an absurd fantasy, and the correlative concept of "nature's own language" is clearly a metaphor' (Kitcher, 2001, p. 46).

In this – epistemological – meaning, ecosystems are clearly not 'discovered', but constructed: they are neither natural kinds, nor is the ecosystem a robust concept. They are thus defined – constructed – by us as a class of objects. This does not deny in any way the physical reality of organisms and their interrelations.

There are always many ways to classify things in nature. In a very intriguing metaphor, Kitcher (2001) uses the image of a chunk of marble, used by a sculptor, to explain this multitude of possible classifications, while still acknowledging that these classifications refer (in different ways) to reality.

Were we to think of the marble as completely continuous, we would see it as containing an uncountable infinity of objects (many of which overlap).

The marble is a little piece of the world and like the large cosmos it can be conceived as divided up into objects in many different ways. Independently of our conceptions, those objects, those chunks of marble, exist. We draw (or chisel) the lines, but we don't bring the chunks into being. There is no determinate answer to the question, 'How many things are there?' and no possibility of envisaging a complete inventory of nature. *(Kitcher, 2001, pp. 44f.)*

In just this way we should also think about ecosystems. This is valid even when we acknowledge that there is not a complete continuity of patterns and processes within the biosphere and that there may be boundaries which – under specific questions! – we consider as more significant than others. Without that, however, we would almost be lost completely when it comes to classifying the biosphere into communities or ecosystems.

Another useful comparison of how ecosystems and their scientific perception relate to reality may be that of a painting (Fig. 5.1): the 'same' chunk of nature will look completely different if painted either by an impressionist, expressionist, cubist, surrealist, or a painter of another style.

Fig. 5.1. Our representations of nature must always select and abstract from reality. Cartoon by Vladimir Rencin; printed with permission of the author.

Yet the very object remains the same, even though there is no way to know it in totality or even its 'very essence'. The painting (the concept) *is not* what it displays (reality), but it is also not completely independent from it. Science means to *select* from the confusing mass of impressions and signals we derive from the world. We as scientists (have to) decide what is important to our specific questions and what is (most likely) not. There is 'reality', but there are always different ways to perceive it. And even more, different *meanings* can be given to the same part of nature.

Not every definition of an ecosystem and of ecosystem functioning, however, is of equal quality and usefulness. Criteria for a sound concept of an ecosystem (or any other scientific concept) are basic criteria such as logical consistency (freedom from contradictions), its relation to other concepts and theories (it should not be conceptually isolated), clarity (being unambiguous), simplicity, and explicitness of assumptions. I have already mentioned one of the most important criteria, especially in applied circumstances – its significance and practical applicability. For further elaboration on the issue of concept formation and application, see Hempel (1952), Essler (1982), van der Steen (1990), Pickett *et al.* (1994), and − more specifically related to concepts of ecological units − Jax (2002a, 2006, 2007).

5.1.3 Does the variety of ecosystem concepts promote environmental relativism?

A fear that arises if nature − and even more if scientific concepts like the ecosystem and its reference states (and thus ecosystem functioning) − are not clearly fixed and predetermined by nature is that of a complete relativity. Relativity means that we may lose firm ground with respect to our perceptions and our knowledge about how the world 'really' is, undermining the credibility of science and scientists. In addition, it refers to moral, and thus political, relativity − summarised here for the sake of simplicity as 'environmental relativism'. The latter amounts to the allegation that environmental problems are not taken seriously, that they are just described as attitudes instead of facts, and that they thus further anti-environmental tendencies ('Why worry when climate change and biodiversity loss are just social constructs?'). Even more than the episte-mological problems described above, this issue has raised much concern and even rage against constructivists' accounts of nature. The problem has been discussed in several places and with some polemics. Opposition has not only been expressed to *sociological* perspectives on the matter, but also to 'relativistic' *ecological* concepts and theories. Thus environmental ethicist Baird Callicott complained about what he called 'deconstructive ecology' (Callicott, 1996). By this he meant the change from the notion of equilibrium ecology to non-equilibrium ecology, which took place in the 1980s. He argued that this new ecology undermined the reference states that conservationists envisioned for ecosystems and thus was also in danger of undermining the scientific basis of some concepts of environmental ethics, in particular that of Aldo Leopold's *Land Ethic* (Leopold, 1949). An argument along the same lines was brought forth in the mid 1980s by a prominent German ecologist, at that time minister for the environment of a German state. During a discussion about the goals of conservation in Germany, he argued that any deviation from the received equilibrium view of ecosystems (or more broadly, the 'balance of nature') − would lead to confusion and a loss of confidence in ecologists among the population. As a consequence, following the further reasoning, this might lead to decreasing public support for biological conservation and environmental protection policies. In a similar sense, arguments were put forth in American conservation biology circles in response, not the least, to William Cronon's provocative book, *Uncommon Ground* (Cronon, 1995), which challenged established ideas of nature and wilderness. Soulé and Lease (1995), in their famous statement, described social constructivism (or deconstructivism as they call it) as a threat to nature, stating that 'certain contemporary forms of intellectual and social relativism can be just as destructive to nature as bulldozers and chain saws' (Soulé and Lease, 1995, p. xvi). They are also convinced that questioning the concepts of nature and wilderness was 'sometimes [used] in order to justify further exploitative tinkering with what little remains of wilderness' (Soulé and Lease, 1995, p. xv).

These are grave accusations. Are they justified? In a certain manner, the way we define ecosystems and ecosystem functioning is really relative – relative to the questions at hand. But that does not mean that there is a complete idiosyncrasy in the sense of a specific definition for each single empirical case (in Section 7.1 I will describe some major types of definition that are commonly used and which are useful to conceptualise ecosystem functioning in the context of ecology and conservation biology). Nor does the fact that ecosystem functioning is relative to specific questions render environmental problems as purely relative, as some authors fear.

In terms of dealing with environmental problems, the idea that nature and ecosystems are partly social constructions does not, in fact, imply *moral* relativity, as many authors ascribe to constructivist approaches. David Kidner, for example, states: 'Constructionism therefore implies a relativistic stance within which one attitude toward or interpretation of the natural world is no better or worse than any other. [. . .] In other words, nature is part of a discursive world, and any "problems" which might exist within this world are produced and solved by debate rather than by embodied action' (Kidner, 2000, pp. 340f.; see also Proctor, 1998 for a discussion of relativist accusation towards 'social constructionists').

Environmental problems are human problems, not problems of nature. The phenomena within nature (e.g. climate change or species exctinctions) that we *perceive* as problems are real, but it is only via human judgements that they appear as *problems*, that they *become* problems. In this way environmental problems are, by necessity, constructed, both physically and conceptually. But it would, as several authors (e.g. Radder, 1992; Gandy, 1996) have pointed out, be logical nonsense to insist that environmental problems are *mere constructs* (also in an ontological sense) and exclusively exist as matters of language and discourse. Otherwise the ozone hole, climate change, or biodiversity loss would physically disappear once we change the discourse about it, without any further physical action needed.

What we observe and judge as environmental problems are the consequence of changes brought about by humans to themselves or to other humans *via* the environment: effects on our health, economy, our aesthetic and moral sentiments, and effects on the parts of the non–human world which we need and/or value (Bird, 1987). And it is just these affected *needs and values* that make changes in the environment an environmental problem, regardless of the 'natural state' of ecosystems, which is, as we have seen, not apt as guidance. But that does not diminish in any way the urgency of problems.

Beyond that, there is another important reason why the multitude of possible and useful definitions of an ecosystem (and with it ecosystem functioning) are not leading to moral relativity. It is in fact not ecosystems 'as such' that we aim to protect, but the elements (e.g. species) and processes (e.g. ecosystem services) of these ecosystems that we value. Ecosystems are conceptions that draw together phenomena, according to these very interests. Our conceptions of ecosystems, if considered consciously, are just preparing us to analyse nature in a way that *allows* us to protect or manage the phenomena (elements and processes) we value.

In an ideal world all the different ecosystem concepts would coincide, they would refer to some natural kind of objects, such as coinciding boundaries. In reality, however, this is not the case. The different processes and objects rarely coincide and thus what is a functioning ecosystem for one person and his or her definition may or may not be a functioning ecosystem for another. The Yellowstone example in Section 4.3 gives evidence for such differences. Likewise, an agricultural ecosystem may be considered as well-functioning by a person who emphasises the delivery of ecosystem services such as those to be used for food or bioenergy, while it may not be a functioning ecosystem for someone who considers an ecosystem to be self-sustained and existing without human intervention. That is, *value judgements* are necessary to decide as to whether a specific patch of the world can be considered a functioning ecosystem (or an ecosystem at all). I will come back to this issue in Section 5.2.

Before diving deeper into the realm of values and societal choices, however, let us take a closer look at what it means in practice that nature is, on the one hand, partly constructed, but on the other hand, resists completely arbitrary classifications. The example of 'alternative stable states' of ecosystems provides an interesting example here. It can help us to think further about the epistemological question of the interplay between nature and the concepts we make of it – in this case in terms of ecosystems and their functioning.

5.1.4 Case study: alternative stable states as distinct modes of ecosystem functioning

Alternative stable states are an interesting phenomenon. It allows us to see that even though the determination of ecosystem functioning is not given by nature, it is still also not only in the eye of the beholder. Some types of ecological systems exhibit strong discontinuities in the performance of their processes, the ensuing patterns, and – which may be the main reason we are aware of it – their physiognomic appearances. The classical and almost paradigmatic of such systems are savannahs, with their transitions to either grassland or forest (van Langevelde *et al.*, 2003); coral reefs, with coral-dominated and algal-dominated states (Knowlton, 1992; Mumby *et al.*, 2007); and shallow lakes with a transition between a clear-water state and a turbid-water state (Scheffer *et al.*, 1993). Alternative stable states have also been described for rivers (Dent *et al.*, 2002), temperate shelf ecosystems (Choi *et al.*, 2004), and other marine systems (de Young *et al.*, 2008). In the following, I will use the example of shallow lakes to explain the phenomenon of alternative stable states and its significance for the issue of defining ecosystem functioning.

Shallow lakes and their alternative stable states
Lakes are ecological systems whose boundaries are (or at least appear to be) very clearly limited. As such, it is no wonder that the first study that ever explicitly applied the ecosystem concept in practice was done on a lake, namely Raymond Lindeman's investigation of the trophic relations in a senescent lake in Minnesota (Lindeman, 1942). Lakes are among the classical objects of whole-ecosystem studies. Scientific investigations of lakes, perceiving them as integrated 'wholes' of biological communities and their environment, were conducted before the word 'ecosystem' was coined (e.g. Thienemann, 1923, 1925). The processes within lake ecosystems are quite well understood.

Lakes can be very different. A prominent distinction can be noted between nutrient-poor, clear lakes with low productivity (oligotrophic lakes), and nutrient-rich, turbid lakes with high productivity (eutrophic lakes). The tendency of a lake to be oligotrophic or eutrophic is dependent on many factors. The most important of these are the morphology of the lake (especially its depth), the size of its catchment, the nutrient input it receives, and its age and history. The same region, with the same climate, may show completely different lake types depending on these variables, as exemplified by the well-studied volcanic Eifel Maar lakes in Germany, where oligotrophic, eutrophic, and mesotrophic lakes are located in close proximity (Scharf and Björk, 1992). But the trophic status of a lake is not fixed. Like anything in nature, lakes are subject to change. They change through the influence of geological processes (e.g. sediment and mineral inputs) and internal biological processes, leading to slow silting up and slow increases in the productivity of the lakes. They also change – much faster – through the impact of human activities. Among the latter, the input of nutrients and its consequences for lake ecology have been the topic of much research (Smith and Schindler, 2009). High nutrient input into a lake, for example, phosphorus and nitrogen from human sewage or fertiliser washed out from soils, leads to complex processes in the interactions between biotic and abiotic components in a lake ecosystem, a process known as (cultural) eutrophication. Nutrient limitation of productivity (characterising oligotrophic waters) vanishes, plant growth (especially phytoplankton growth) is enhanced, and the decomposition of the plant biomass finally leads to oxygen depletion in the sediment and large parts of the water body.

But while the growing amount of nutrients leads to rather gradual change of biota and chemistry in many lakes, some lakes exhibit a rapid and almost abrupt transition from a clear-water stage to a turbid-water stage. Once a *threshold* in the level of nutrients is crossed, the lake shifts to an *alternative stable state*. This phenomenon is especially pronounced in shallow lakes (Scheffer *et al.*, 1993). In shallow lakes, sunlight can, in principle, penetrate to the ground, enabling macrophyte growth in large areas of the lake bottom. The macrophytes allow zooplankton organisms to escape predation by fish, and by their own growth the plants assimilate nutrients. They also diminish bioturbation by fish and dampen its effects (turbidity through suspended fine sediments). Low nutrient levels (especially phosphorus) and zooplankton keep phytoplankton growth and density at low levels. All this keeps the lake in a *clear-water state*. With increasing nutrient levels, however, phytoplankton growth accelerates and increasingly limits the light available for macrophyte growth, finally leading to a non-linear sudden decrease of the macrophytes. In consequence, zooplankton is further diminished (loss of hiding places, increased fish predation), allowing even further phytoplankton growth. In addition, bioturbation by fish increases, and the fine sediments thus increasingly suspended further diminish the light that is able to penetrate to the ground. The lake has flipped to a *turbid-water state*. This new state will not simply be reversed if nutrients are reduced to the level below the threshold where the original change occurred. Only at much lower nutrient levels than those at which the transition occurred can the lake be reversed again to a macrophyte-dominated, clear-water state. This phenomenon is known as *hysteresis* (see Box 5.2). One reason for this is that excess nutrients that have been 'trapped' in the sediment can be easily remobilised when oxygen levels are sufficient and keep the production of phytoplankton going even when external nutrient loading is diminished.

Box 5.2 *Types of thresholds in ecosystem dynamics*

Ecosystems respond to external changes in different ways (Fig. 5.2). If the external driver (e.g. nutrient input or temperature) changes gradually, the ecosystem state may respond gradually itself (Fig. 5.2a), by a slow change in relative species abundances or a change in productivity, for example. In this case, there are no thresholds involved for the properties observed that characterise the ecosystem state. In other cases, the response is abrupt (Figs. 5.2b–5.2c), even though the external variable changes only gradually. While there is little response below a certain threshold value, a small change in the environmental condition will lead to a fast change of the system state once that threshold is crossed. Cases b and c, however, are decisively different. The system state in b (as well as in a) are easily reversible, in the sense that the way back to the former state is the same as that to the new states (reducing the magnitude of the driver below the threshold value). In case c, however, the change is either irreversible or at least not reversible using the original path. It is completely irreversible if there is no return to the original system at all, even if the external change is completely reversed. This may be the case when crucial species have gone extinct, or when parts of the environment are altered – by erosion, for example. In other cases, reversal to the first state is possible, but not easy. Here, hysteresis in the system requires that the environmental factor that triggered the change be set back to much lower levels than that of the threshold for the 'forward journey'. There are different thresholds for system transitions in the different directions ($F1$ and $F2$ in Fig. 5.2c). Within the range of environmental

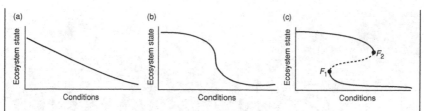

Fig. 5.2. Different types of ecosystem responses to changing external conditions. See text for explanation. Graphics reprinted from Scheffer and Carpenter (2003, p. 649). Copyright 2003, Elsevier, and 2001, *Nature*, reprinted with permission.

conditions between these two thresholds (i.e. under the same environmental conditions), the system can persist in either state.

In general, all sudden and severe system shifts are characterised as *regime shifts*. Andersen *et al.* (2009, p. 50) define an ecological regime shift as 'a sudden shift in ecosystem status caused by passing a threshold where core ecosystem functions, structures, and processes are fundamentally changed'. Only shifts which involve hysteresis, however – and thus the possibility of a system existing in two different states (or regimes) at a range of *identical* environmental conditions – are called shifts between *alternative stable states* (or alternative stable regimes).

One of the lakes in which the phenomenon of shifts between alternative stable states was investigated is Lake Veluwe, a shallow lake in the Netherlands (van der Molen, 1998; Ibelings *et al.*, 2007). It was created artificially in 1956 and has a mean depth of only 1.5 m. In its first years it was a clear-water lake with low nutrient levels, dominated by macrophytes, especially *Characeae*. Anthropogenic nutrient inputs strongly increased during the late 1960s, leading to a shift towards a turbid state, with no *Characeae* left, a dominance of phytoplankton (especially large blooms of the cyanobacterium *Planktothrix agardhii*), and extremely low transparency (less than 20 cm of Secchi-disk depth) during the 1970s and 1980s. Restoration efforts, beginning in the late 1970s, were able to revert the lake to a clear-water state again. Restoration measures consisted in the first place of strongly reducing the phosphorus reaching the lake and flushing the lake with nutrient-poor and calcium-rich (both favourable to *Characeae*; see Fig. 5.3) water during the winter periods. The return of macrophytes to the lake showed a clear hysteresis pattern. While a significant cover of *Characeae* was still present at a phosphorus

Fig. 5.3. Plant of the genus *Chara*. *Characeae* are multicelluar branched algae. They often form underwater meadows, with some species reaching sizes up to 1 m. Photo courtesy of Stefan Zimmermann, Limnologische Station Iffeldorf TU München.

concentration (summer mean) of 0.2 mg l^{-1} (total phosphorus, TP) in 1969, the first charophytes only returned when TP had fallen below a level of 0.1 mg l^{-1}, which happened around 1990 (Ibelings *et al.*, 2007). But the decisive aspect is that a threshold behaviour with hysteresis was not only observed for one species group, but for several ecosystem variables, such as turbidity, light attenuation by non-algal components, and zebra mussel density. There are thus strong arguments for the existence of alternative stable states in this system, more so as the processes within each of the states tends to reinforce and perpetuate each state (either clear-water or turbid-water state). In terms of describing the functioning of the lake ecosystem, we may consider the two states as two *modes of lake ecosystem functioning*. It seems that there is, at least in such cases, not a continuum of how organisms and their abiotic environment interact, but a strong discontinuity. It appears as if we can find here some reference state for characterising the functioning of ecosystems. As one of these states (mostly the turbid state) is considered as undesirable from a societal point of view, there are attempts to identify such thresholds for classifying the 'ecological status' of water bodies in the context of the European Water Framework Directive (WFD) (Lyche Solheim *et al.*, 2008). Can we here find a way of defining ecosystems and their functioning independent from specific observer perspectives?

Ecosystem functioning and regime shifts
In terms of assessing ecosystem functioning, the decisive characteristic of alternative stable states (beyond their hysteresis properties) is the following: the transition between the states is abrupt, and several systems variables ('multiple aspects of physical and biological components' (Lees *et al.*, 2006, p. 106)) change concurrently. Also, the respective state of the system is maintained under a range of external changes. However, a few questions remain, and these show that even a stable state (or regime; see below) of the kind described above does not offer unambiguous reference states for ecosystem functioning.

First, finding conclusive evidence for the existence of stable alternative states in specific ecosystems is not easy (Scheffer and Carpenter, 2003), not the least because, as Beisner *et al.* (2003, p. 380) state: 'Ecologists and philosophers of science have not yet agreed on how different a state must be in order to be deemed truly alternate.' There is also a general complaint about the scarcity of good empirical, especially experimental, evidence (Schröder *et al.*, 2005). Even Ibelings *et al.* (2007, p. 12) see their results from Lake Veluwe and its transitions between clear-water states and turbid-water states only as 'hints' towards the existence of alternative stable states. There are, therefore, many discussions about the usefulness of the concept with respect to how, where, and under what circumstances we can really empirically assess and predict the existence of multiple stable states and the thresholds for their transitions (Beisner *et al.*, 2003; deYoung *et al.*, 2008; Andersen *et al.*, 2009).

Major questions that remain here are: what are the variables by which we assess ecosystem functioning, or in this case, by which we distinguish between different modes of functioning? Is there a completely different set of variables in each type of ecosystem, or are there common elements and processes? How important are physiognomic aspects of the ecosystem as compared to other pattern-related and process-related variables? How different do the systems have to be to be called alternative stable states? Answers to these questions vary. I think that the notion of alternative stable states today is linked mostly to an idea of ecosystems that is far more abstract than the approaches followed in most traditional fields of conservation. By this I mean that researchers perceive ecosystems not as composed of specific species compositions or narrow reference intervals of process rates, but as composed of rather coarsely resolved components and processes. This is mirrored also in the terminology used. Researchers dealing with alternative stable states (and/or ecological resilience) often try to avoid the word 'state' as it connotes a 'static' condition of a system for many readers. Instead of 'state' they use the terms 'configuration' (Walker *et al.*, 2002), 'attractor', or – more frequently – 'regime' (Scheffer and Carpenter, 2003) to describe the dynamic condition of an ecosystem, including fluctuations and trends. Such a regime or attractor is seen as a 'collection of states' (Walker *et al.*, 2002, p. 14), which might by other definitions even be characterised as a collection of different ecosystems. But even if we acknowledge (and I think we must!) that ecosystems are highly dynamic systems, there is also always a constancy of *something* through time (self-identity; Jax *et al.*, 1998). Otherwise we would not be able to speak of one and the same 'system' (and its regime) in two instances of time at all. Even the most 'statically' defined ecosystem (e.g. by specific species compositions and species abundances) is dynamic because every organism in it exchanges matter and energy, is born, moves, dies, and often also reproduces. And even the most dynamic system definition requires some variables that are considered to be constant. Rejecting equilibrium ideas of ecological systems, Holling, in his seminal paper on resilience, states: 'But if we are dealing with a system profoundly affected by changes external to it, and continually confronted by the unexpected, the constancy of its behavior becomes less important than the *persistence* of the relationships' (Holling, 1973, p. 1; my emphasis).

It is this relationship, the 'regime' or 'configuration' which is thought of as being constant. The perspective is just shifted to another, more abstract perspective and level of description than before. Or, as Walker and colleagues put it: 'We are usually interested in preserving a particular set of *general* criteria' (Walker *et al.*, 2002, p. 5; my emphasis). So the question of what a functioning ecosystem is, is not simply solved by referring to regime shifts and alternative stable states. The detection of thresholds can nevertheless be very helpful in classifying systems of a particular type of ecosystem (definition).

Questions to be answered are: how can we pinpoint or even predict thresholds for the different states of the ecosystem (i.e. different modes of its functioning)? How variable are these thresholds? Which parameters are most sensitive? The literature in search of thresholds shows that there are species that are highly sensitive to particular changes (e.g. nutrients), while others are insensitive (Lyche Solheim *et al.*, 2008). Also, the precise location of thresholds (also called 'tipping points') is not always fixed to the same value of a variable (e.g. nutrient content) (Carpenter and Lathrop, 2008).

The example of alternative stable states can teach us several things: first, for some kinds of ecological systems, there may be discontinuities in the way major processes and patterns operate, instead of gradual shifts. Some people may perceive the different, 'disconnected' states as privileged reference states. However, not all systems exhibit such alternative states, with a large number of observation variables changing simultaneously and in a non-linear fashion (Schröder *et al.*, 2005). It also remains unclear which kinds of systems exhibit alternative stable states and which do not, although there are some hypotheses about it (e.g. Didham *et al.*, 2005).

Second, the question of what constitutes alternative stable states (or alternative stable regimes) is both dependent on physical reality and on the specific *selection* of variables an observer sees as decisive for characterising an ecosystem and its functioning. It is interesting to see that most of the examples given for regime shifts are characterised by specific *physiognomic* shifts, i.e. shifts in the dominant life forms of a system. This suggests a prominence of variables that relate to physiognomy of ecological systems in such approaches.

The responses of organisms, their interactions with each other and their abiotic environment, are clearly independent of human observers. The question of what kind of dynamics we, as observers, include or exclude as part of the (internal) functioning of an 'ecosystem' (i.e. the very definition of an ecosystem), however, is likewise important for assessing alternative stable states. It decides to a considerable degree whether we interpret discontinuous shifts in related ecosystem variables (1) as merely a variation in ecosystem functioning; (2) as different modes of functioning (alternative states); or (3) as a change of one ecosystem into another (functioning) ecosystem. The search for ecosystem thresholds is nevertheless highly useful because it helps, if successful, to restrict the range of possible reference states for ecosystem functioning. Alternative stable states do not allow us to characterise ecosystems as 'natural kinds', but they can help provide reference conditions for some conceptualisations of ecosystems.

Finally, even if we agree to a common definition of the ecosystem, and if all ecosystems had clearly delimitable alternative stable states, the question of which of these alternative states is *preferable* (or desirable) is still a matter of evaluation by human observers. Each alternative state, or expressed differently, each mode of functioning, can be considered as 'proper functioning' for its kind of system and state. Each transition has 'winners' and 'losers' among the organisms involved and impacted – and also in terms of the services humans may derive from the system. Although a highly eutrophic lake may be a highly persistent and properly functioning (and even resilient) ecosystem, we may not *want* it to be that way, we may value it as being something undesirable.

5.2 Values, norms, and ecosystem functioning: a necessary and difficult unity

Conceptualising ecosystem functioning is a matter of both scientific knowledge and values. According to the classical philosophy of science (and of the self-image of many natural scientists), however, such concepts are 'forbidden', because science should be value-neutral. The received view, in the tradition of Max Weber, states that the empirical sciences must be free from values and normative dimensions. Yet, hybrid concepts, embracing both descriptive and normative dimensions, *exist* in ecology, and especially in conservation biology – and they proliferate (Callicott *et al.*, 1999). The normative character is quite evident in concepts like 'ecosystem services', 'ecological integrity', 'ecosystem health', or 'ecological restoration'. It is less evident, at least at first sight, in the concept of 'ecosystem functioning' or some other much-used scientific concepts such as 'biodiversity' or 'resilience' (Brand and Jax, 2007). All these concepts are what has been called 'epistemic–moral hybrids' (Potthast, 2000a, 2005).

The postulate that science must be (or *is*) 'value-neutral' has been challenged repeatedly, from different perspectives. On the one hand there has been the promise or postulate that, specifically, ecology was a model for a new kind of science, a science that closes the hiatus between facts and values, between is and ought, that is descriptive *and* normative (e.g. within the deep ecology movement, see Naess, 1973; see also Trepl, 1983). This is a very problematic postulate, because it suggests that values and norms can be taken directly from nature. In such a science, nature itself would provide norms for human actions, and the task of ecologists would, in consequence, be to *discover* these norms through their descriptions of how nature is. However, to deduce normative statements (an 'ought') *directly* from descriptive statements (an 'is') means to commit what is often called a *naturalistic fallacy* (e.g.: *because* the natural state of a forest *is* such and such, we *have to* protect it in that state). Expressed another way, committing a naturalistic fallacy would be to say that what is natural is *automatically* also good. But nature does not provide moral norms. It is always us, humans, who set norms and value things. The fact that sheep do not produce red wool does not imply that we are not allowed to wear red wool sweaters, and the fact that there is 'naturally' infanticide among primates does not make infanticide morally legitimate for us. This does not mean, on the other hand, that it would not be legitimate to protect a 'natural' state of an ecosystem or to follow nature in our lifestyles, and even orient norms towards such goals. But the goal cannot be derived *directly* from nature. Instead, a value-based decision – on the basis of good arguments – is required to connect factual information to statements about what we *should* do.

Criticism of the thesis that science is completely value-neutral (at least in the simple form often presented by scientists) is, however, justified from another angle. To some degree, value decisions are unavoidable, including within science. Even at the level of pure science, every act of selecting a research question, selecting observation variables, selecting theory and concepts (and the assumptions they are based on), as well as selecting adequate ways of interpreting data, are, at least implicitly, value-based decisions (Tucker, 1979). However, such values are not necessarily *morally* relevant values. There are different kinds (or degrees) of normativity. First, there are, by necessity, values and norms relating to why and how we select and decide upon the objects of our research (what we value as interesting, important, and relevant), and what good conduct of scientific methodology is (e.g. which methods lead to desired results, are reproducible, precise, etc.). These have been called 'constitutive values' by Longino (1983) or 'methodological values' by Shrader-Frechette and McCoy (1993). Max Weber (1917 [1968]) acknowledged the existence of such values, but they were of no concern to his thesis of a value-neutral science, were not what he meant by values from which science should keep clear. For him, being value-free in the first place meant being free from *moral* values (Stegmüller, 1986). Nevertheless, methodological values must be dealt with consciously and should not be taken for granted. They strongly influence how we conduct science, how we construe our concepts, what we perceive as significant or insignificant, important or unimportant. Moreover, what we consider as important is largely influenced by societal contexts. It is influenced by societal (morally relevant) value systems, by scientific fashions, changing theoretical assumptions, and by personal interests (e.g. in terms of allowing publications in highly rated journals). The boundary between methodological values (internal to science) and societal values (external to it) are thus blurred significantly (see case studies in Longino, 1983). As has been discussed above, the reason biodiversity and ecosystem functioning are investigated is to a considerable extent determined by a conservation concern, with a variety of different values implied. The other way round, our research has an influence on society's value systems, partly unintentionally, partly intended. Through their research on ecosystem services, scientists have raised the appreciation of the values of biodiversity and ecosystems. Influences such as these can become morally highly relevant, not the least because different research fields compete for financial resources and because research in conservation biology may lead to political actions affecting many people's lives.

There are other values embedded in our scientific concepts and practices with a much stronger societal dependence than purely methodological values. These values do not refer to the *mode* of doing science, but towards evaluating the *contents* it describes. They imply a stronger normativity. Such values are at stake when assessing and even conceptualising 'ecosystem functioning', especially when it comes to defining what a 'functioning ecosystem' is. Doing this, by necessity, involves value decisions: we are ascribing *norms* to the performance of the system (*proper* performance), norms which nature cannot provide to us, but which have to be taken from societal contexts and choices. While such choices are unavoidable when dealing with ecosystem functioning, there is a gradient in terms of the moral dimensions of these norms. The moral dimension increases the more we move into applied fields, such as conservation biology.

Conservation biology as a science is normative by definition (Soulé, 1985). 'Conservation' is in itself a value-laden term, formulating a specific societal goal in terms of supporting a specific direction of human actions, i.e. nature conservation. Research in conservation biology is oriented towards supporting these goals. It is hardly conceivable to establish a science of 'destruction ecology', at least not with the goal of promoting destruction. There is, however, considerable disagreement about how far its normative perspective should extend, in terms of advocacy, for example (e.g. Barry and Oelschlaeger, 1995; McCoy, 1996; Meine and Meffe, 1996; Shrader-Frechette, 1996).

As such, normative concepts should, in principle, be less of a problem for conservation biology, but much more for ecology, which generally is considered as a classical value-free science by most of its practitioners. With respect to the notion of 'ecosystem functioning', the term appears much more neutral than related terms such as 'ecosystem integrity' or 'ecosystem health'. But in fact, the difference is at best gradual. Where 'health' and 'integrity' allude directly to the proper performance of organisms and/or social institutions, 'functioning' alludes more to the performance of artefacts (we would rather speak of the proper functioning of a car than of its health and integrity). 'Functioning', as well as the term 'ecosystem services', at least from the metaphorical connotations they carry, are more open to the fact that the norm which is included is not set by nature as such – as 'health' suggests – but is based on individual and/or societal interests.

So should we abandon the concept of 'ecosystem functioning' in ecology because it relates to values and embraces normative aspects? The answer is 'no'. What appears as a problem may be turned into an advantage, provided that some preconditions are satisfied. The problem that 'ecosystem' and the notion of 'functioning' are observer dependent can be approached much more easily when acknowledging the normative dimension of the concepts – and vice versa. In fact, it provides part of the solution for conceptualising and operationalising ecosystem functioning. There is not one fixed norm in nature for determining ecosystem functioning. Instead there are different possibilities of what a functioning ecosystem may be (see Chapters 6 and 7). Only by relating societal values and choices with research on ecosystems can we assess their functioning. It is not the normative dimensions of concepts as such that cause problems. Problems arise if and when the normative aspects are not made *explicit*, but disguised as purely objective and scientific (Pielke, 2007). Scientific credibility and objectivity is not diminished by acknowledging the existence of values in our concepts (Barry and Oelschlaeger, 1995), as long as we are explicit about them. To the contrary: disclosing the values implied in specific definitions of concepts like 'ecosystem functioning' allows us to discuss these values and thus make them open for societal discourse. The classical postulate of Max Weber and others remains valid and becomes even more important: to *distinguish* clearly between descriptive statements and evaluations. This is often difficult, because in the practice of conservation, and sometimes even within research, both dimensions are closely intertwined. But from the perspective of sound science and philosophy, as well as from that of social responsibility, it is indispensable. Only when we first distinguish descriptive and normative aspects – and discuss both in their own right and at their respective level – can we put them together again in a sound manner for the practice of research and application.

5.3 Case study: societal choices and ecosystem management – the Ecosystem Approach of the Convention on Biodiversity

If ecosystem functioning is situated between empirical reality and societal choices, how do we deal with it in practice? This is especially important when ecosystem functioning concepts are applied in the context of environmental management. One such field, in which ecosystem functioning has become an increasingly important concept, is that of *ecosystem management*. In particular, one of its strategies, expressed in the so-called 'Ecosystem Approach' of the CBD, can serve as a good example of the challenges that arise when science and society meet with the aim of ensuring the continued functioning of ecosystems.

5.3.1 The development of ecosystem management approaches

Since its early days, such as in Yellowstone National Park (Section 4.3), ecosystem management has developed from a purely ecology-based approach into a highly interdisciplinary and even transdisciplinary one. At first the most important and then revolutionary shift was that from managing specific selected species or natural features towards managing whole ecosystems. This involved an emphasis on ecological (i.e. 'natural') boundaries for management areas instead of administrative ones. This, in turn, necessitated interagency cooperation in management, which soon became a second hallmark of ecosystem management. A further decisive element of ecosystem management today is inclusion of stakeholders and different knowledge forms beyond scientific knowledge (e.g. traditional and local knowledge) in planning and implementation. A huge number of concepts and projects on ecosystem management have been developed and applied, especially in the USA and Canada. While concept and practice of ecosystem management are highly heterogeneous in detail, the overall philosophy has converged during the 1990s in terms of the characteristics described above. Overviews on the philosophy, methodology, and practice of ecosystem management can be found in Grumbine (1994b), Christensen *et al.* (1996), Boyce and Haney (1997), and Lackey (1998).

Ecosystem management has reached the political arena both in national policies (e.g. the establishment of the Federal Ecosystem Management Initiative in the USA during the 1990s (Malone, 2000)), as well as in international contexts. On the international level, ecosystem approaches have been codified as a preferred mode of management by several

Box 5.3 *The Convention on Biodiversity*

The CBD was adopted by the World Congress on Conservation and Development (UNCTAD) in Rio de Janeiro on 5 June 1992. It came into force on 29 December 1993. It currently (September 2009) has been signed by 191 parties, 190 nation states, and the European Community. The only states who (to this date) have not become parties are Andorra, the Holy See, Somalia, and the USA. The CBD is an ongoing process with a complex array of institutions and a funding mechanism, in the form of the Global Environmental Facility (GEF). Its main decision-making body is the Conference of the Parties (COP), which meets every two years. For further detailed information about the CBD, its contents, processes and institutions, see its website: www. cbd.int. The CBD and its EA are not legally enforceable, but constitute what is sometimes called 'soft law'. That is, its implementation depends on the voluntary commitment of the parties. In spite of this, it has a considerable impact on global and national conservation policies.

international programmes, such as those of IUCN, UNEP, UNESCO's Man and the Biosphere Programme, and the CBD. Working groups and commissions have been formed for this purpose, groups that developed a number of ecosystem management strategies and guidelines. The general ideas of ecosystem management have nowhere been expressed as succinctly as in what is called the Ecosystem Approach (EA) of the CBD. In the following, I will use the text of the EA to characterise the role of ecosystem functioning within ecosystem management approaches and to discuss the tensions between the roles of science and society inherent in these strategies.

5.3.2 The Ecosystem Approach of the Convention on Biodiversity

While at first largely unrelated to ecosystem management, the development and adoption of the CBD was a pivotal event for establishing a new dimension of interaction between science and society in connection with biological conservation. In particular, it has extended conservation far beyond its traditional domain by linking it with the use of biodiversity and the equitable sharing of benefits deriving from this use. This triad of aims (conservation of biodiversity, its sustainable use, and the equitable sharing of benefits) is the core of the CBD. The history of the CBD negotiations (McConnell, 1996) shows how the original conservation focus was increasingly complemented by economic and social concerns, thus also bringing into contact societal groups which hitherto had been strongy opposed to conservation issues. Also, as described in Section 3.1, 'biodiversity' was, from the beginning, introduced as a topic of societal relevance.

One of the major tools for implementing the CBD is its so-called Ecosystem Approach. The EA is understood as 'a strategy for management of land, water and living resources that promotes conservation and sustainable use in an equitable way' (UNEP/CBD, 2000, pp. 103f.). It was approved by the 5th COP of the CBD in Nairobi in 2000 (Resolution COP V/6; see Hartje et al., 2003 for a more detailed history). It is an instrument for the implementation of the three goals of the CBD. The 12 so-called 'Malawi Principles' (see Box 5.4) describe the main elements of the approach. The principles were supplemented by a short rationale for each of them and by five points of 'operational guideline' for the application of the EA (UNEP/CBD, 2000).

The Malawi Principles postulate a close connection between scientific and societal aspects of biodiversity conservation. On the one hand, they refer to the language, the authority, and the information basis of science, but at the same time set science and its results into a broader societal framework. Very prominently, Principle 1 states: 'The objectives of management of land, water and living resources are a matter of societal choice.' That is, determining the objectives of management is not a matter of science or conservation biology alone. This may be trivial to some people, but in fact, biological conservation is still often considered mainly as a matter for experts and as being largely determined by science. In the following principles the inclusion of the economic context (Principle 4) and the use of all relevant knowledge (including traditional knowledge; Principle 11), as well as the involvement of 'all relevant sectors of society and scientific disciplines' (Principle 12) is demanded, emphasising the social dimensions of ecosystem management further. On the other hand, the use of sound science and modern ecological theory is seen as absolutely necessary for implementing the EA. Thus, theories of ecosystem functioning (Principles 5 and 6), scale (Principles 7 and 8), and non-equilibrium dynamics and disturbance ('change is inevitable'; Principle 9) are referred to. The EA is replete with ecological concepts. Many of these concepts are highly complex and ambiguous and it is far from obvious how to apply them in a specific research and application context.

Box 5.4 *The Malawi Principles*

- *Principle 1*: The objectives of management of land, water and living resources are a matter of societal choice.
- *Principle 2*: Management should be decentralised to the lowest appropriate level.
- *Principle 3*: Ecosystem managers should consider the effects (actual or potential) of their activities on adjacent and other ecosystems.
- *Principle 4*: Recognising potential gains from management, there is a need to understand the ecosystem in an economic context. Any ecosystem management programme should:
  (a) reduce those market distortions that adversely affect biological diversity;
  (b) align incentives to promote biodiversity conservation and sustainable use; and
  (c) internalise costs and benefits in the given ecosystem to the extent feasible.
- *Principle 5*: Conservation of ecosystem structure and functioning, in order to maintain ecosystem services, should be a priority target of the ecosystem approach.
- *Principle 6*: Ecosystems must be managed within the limits of their functioning.
- *Principle 7*: The ecosystem approach should be undertaken at the appropriate spatial and temporal scales.
- *Principle 8*: Recognising the varying temporal scales and lag-effects that characterise ecosystem processes, objectives for ecosystem management should be set for the long term.
- *Principle 9*: Management must recognise that change is inevitable.
- *Principle 10*: The ecosystem approach should seek the appropriate balance between, and integration of, conservation and use of biological diversity.
- *Principle 11*: The ecosystem approach should consider all forms of relevant information, including scientific and indigenous and local knowledge, innovations and practices.
- *Principle 12*: The ecosystem approach should involve all relevant sectors of society and scientific disciplines.

Source: UNEP/CBD, 2000

Several of the concepts (ecosystem functioning, disturbance, and, of course, biodiversity) are also hybrid concepts as described above.

The EA is thus about much more than simply the 'ecosystem' in the sense of pure ecology. In accordance with the CBD, it instead reaches far into economic and social dimensions. 'Ecosystem functioning' figures as a prominent concept within the EA. Principle 5 calls it a 'priority target' of the EA. It is, however, not defined in much more detail. So the question must be raised as to what ecosystem functioning means and how its 'limits' (Principle 6) can be assessed. Beyond that, one could argue that the ideas about conserving ecosystem functioning (and structure) as formulated in Principle 5 and especially in Principle 6 (to manage ecosystems 'within the limits of their functioning'), contradict the notion of societal choice as expressed in Principle 1. How can we choose management objectives for ecosystems when ecosystem functioning is given by the natural system and when its limits are given by nature? If ecosystem functioning, however, is not determined by nature, how do we come to reference states and make sense of Principles 5 and 6 at all? Is every ecosystem state equally valid? Certainly not. 'Ecosystem' here must refer to *proper* functioning; otherwise the postulates formulated in the principles would make no sense. The overarching question which we must answer here is thus how our scientific understanding of ecosystem functioning relates to the societal choices postulated.

5.3.3 'Ecosystem functioning' in the Convention on Biodiversity Ecosystem Approach

I am fully aware that the Malawi Principles have been formulated explicitly – and with purpose – in a very general way, that they can and must be adopted to specific circumstances of their application. Acknowledging this, it is nevertheless possible and useful to analyse how far the text of the EA and its interpretations can help us to answer the above questions when it comes to implementing the EA.

The CBD (Article 2) defines 'ecosystem' as: 'A dynamic complex of plant, animal and micro-organism communities and their non-living environment interacting as a functional unit.' This definition comes close to the generic meaning of an ecosystem (Chapter 4). It thus provides not much help for answering the question of how the functioning of an ecosystem might be assessed and how it should be managed. There are also no further explicit specifications given in the EA itself. Principle 5, however, at least provides some hints about the intended meaning of 'ecosystem' and 'ecosystem functioning'. It does so by specifying more clearly the *purposes* of protecting ecosystems: ecosystem structure and functioning should be maintained '*in order to* maintain ecosystem services' (my emphasis). Ecosystem functioning here clearly does not refer simply to some (any) processes at the ecosystem level. It refers to the *overall performance* of the system, with the special purpose of allowing for the provision of ecosystem services (and, of course, to protect biological diversity — as the general aim of the CBD). The rationale provided for Principle 5 adds a further specification: 'The conservation and, where appropriate, restoration of these interactions and processes is of greater significance for the long-term maintenance of biological diversity than simply protection of species' (UNEP/CBD, 2000, p. 106). It is therefore in line with the postulates of Brian Walker (1992, 1995), who considers the protection of ecosystem functioning as a means to preserve as much biodiversity as possible, but not so much specific species. That is, the idea of a (properly) functioning ecosystem here is less detailed than that aimed for by the managers of Yellowstone National Park, for example, which aim at protecting *all* species historically present in the park (Section 4.3).

With respect to the 'limits' of the functioning of ecosystems (Principle 6), the respective rationale only specifies that these limits are constrained by the 'environmental conditions' and that they 'may be affected to different degrees by temporary, unpredictable or artificially maintained conditions' (UNEP/CBD, 2000, p. 106). In a guideline to the EA provided by the Secretariat of the CBD (2004, p. 18), the authors further add to the rationale of Principle 6 to explain the 'limits of functioning':

There are limits to the level of demand that can be placed on an ecosystem while maintaining its integrity and capacity to continue providing the goods and services that provide the basis for human wellbeing and environmental sustainability.

This quote emphasises that ecosystem functioning in the EA is aimed towards providing ecosystem services. Beyond that, however, a new term is added, namely that of (ecosystem) *integrity*, possibly as a characteristic of the system independent of specific use-related notions of functioning. However, as is often the case, also in scientific texts in ecology and conservation biology, in this way another complex unexplained term is introduced. It does, in fact, not contribute to clarity, even though many readers will have a vague and intuitive idea of what might be meant by the 'integrity' of an ecosystem.

In another paragraph of the same text (CBD, 2004), it is stated that there are also limits to ecosystems in terms of the amount of *disturbance* they can tolerate. Furthermore, the notion of limits to ecosystem functioning is identified with 'thresholds for change' beyond which 'an ecosystem undergoes substantial change in composition, structure and functioning, usually with a loss of biodiversity and a resulting lower productivity and capacity to process wastes and contaminants' (CBD, 2004, p. 18). This alludes to the concept of regime shifts, as described above.

All in all, the meaning of ecosystem functioning within the EA, although highly biased towards a system characterised by its ability to provide ecosystem services and maintain biodiversity, remains rather elusive. Given that the EA is thought to be applicable to a broad range of ecosystem types (e.g. forests, marine systems, grasslands) and situations (climatic, social, economic, cultural), a very narrow definition might not be appropriate. What would be desirable, however, is some guidance in how to arrive at more precise definitions for assessing ecosystem functioning in specific cases – in terms of a procedure. The guidelines developed up to now are not discussing this issue, nor do they relate the postulate of societal choices (Principle 1) to it. This is not surprising, as normally ecological concepts are not considered subject to 'societal choices' and participative methodologies. Only the specific management aims and the implementation process are. But even if it might not be intended in the EA, my plea is that some of the important ecological concepts, which are crucial ingredients of the EA, should and even must be subject to societal choices, the 'ecosystem' and 'ecosystem functioning' being prominent among them. That is, participation and societal choices are necessary at the planning stages of research projects, not only at their implementation stage. This would really constitute a big step in connecting science and society. It will, however, require translation between a scientific language and other forms of expression, especially everyday language. I will come back to these issues in Chapter 7.

The fact that the EA does not solve the tension between science and society or fact and value in the course of its efforts to protect biodiversity and ensure ecosystem functioning should not be seen as a flaw. To the contrary: what is important is that the EA opens our eyes to this tension at all. It is important that this tension is maintained. It is productive. It forces us to rethink the roles of both science and society in a conservation context. 'Ecosystem functioning' and 'biodiversity', as well as some other

Box 5.5 *Beyond the Ecosystem Approach of the Convention on
Biodiversity: related approaches*

Different ecosystem management strategies of different international
programmes have merged or at least referred to each other in many
respects. The EA of the CBD has been based on earlier ideas formulated by IUCN's Commission on Ecosystem Management (see Hartje
et al., 2003, p. 9). Vice versa, IUCN now promotes the CBD EA in
its publications (e.g. Shepherd, 2004, 2008). Also, UNESCO's Man
and the Biosphere Programme (established in 1970) today emphasises
the compatibility and even complementarity of the EA to the concept
of Biosphere Reserves, promoting it as the appropriate strategy for
their management (UNESCO, 2000; Bouamrane, 2007). Even par-
allels between the EA and the Water Framework Directive (see the
case study in Section 6.2) have been pointed out, though with the
important difference that the management objectives of the WFD
('good ecological status') are rather fixed, in contrast to the postulate
of Principle 1 of the EA that they should be a 'matter of societal
choice' (Nõges *et al.*, 2009). More recently, a new ecosystem management programme was set up by UNEP (UNEP, 2008). Many of
its formulations have been taken explicitly from the EA, merging it
with the general scheme of the Millennium Ecosystem Assessment.

concepts, can build bridges between science and society, precisely *because*
they embrace both descriptive and normative dimensions.

5.4 Conclusions: the roles of science and society in assessing ecosystem functioning

5.4.1 General conclusions

In the case study dealing with the intactness of the Yellowstone ecosystem
(Section 4.3.1), it was shown how a multitude of influences shape the
definitions of what a functioning ecosystem is. These range from specific
ideas about nature to specific interests and rules of administrative institutions through scientific paradigms. As such – and this holds even more
for basic research on ecosystems – 'ecosystem functioning' is determined
not just by aspects of an applied character (such as management interests), but also by choices made by the individual researcher and by the
research community of which he or she is a member. What counts as a

scientifically permissive and interesting question is dependent on the prevailing scientific paradigms and scientific theories, but also on scientific fashions or what is good for individual careers. This, in turn, is shaped in a complex manner by funding agencies, scientific journals, university departments, and the power relations within all these institutions. Doing science is also a societal enterprise and not a purely individual and logical one (e.g. Longino, 1990; Jasanoff, 2004).

All this is, of course, not independent of empirical findings that ecologists make during their work, but neither does empirical research on ecological systems directly and in a clear and unambiguous manner determine how ecosystems are defined (see the case study on alternative stable states of lakes in Section 5.1). In reviewing the different definitions of ecosystems and other ecological units (such as the community) during the course of the history of ecology, I could detect no clear trends in theory development (Jax, 2002a, 2006). That is, there is no indication that some kinds of definitions were completely abandoned in favour of more 'advanced' ones, approaching the 'true' ecosystem definition – even though the amount of knowledge assembled about ecological patterns and processes has increased enormously during the same period. The plurality of ecosystem definitions has persisted and has even been extended over time, which also hints at the variety of influences that actually determine the selection of what an ecosystem is and how its functioning may be characterised. To select the appropriate definition of an ecosystem depends very much on analysing the specific task of research and, even more in applied research, the specific conservation or resource management context. Within these contexts, decisions (societal choices) must be preceded by a clarification of the often implicit ideas about what constitutes a functioning ecosystem – what it means to the scientists as well as to the (other) stakeholder groups involved. Ideally, such clarifications should be the basis of research already (and not only of management), because they determine the kind of relevant data to be assessed.

The role of ecology in the context of new approaches to biological conservation is becoming both restricted and enlarged (see also Chapter 7). It is restricted because its dominance in conservation and resource management is questioned. Ecology (and other natural sciences) cannot simply claim priority over the humanities or other forms of non-scientific knowledge any more, nor can ecological research and knowledge substitute societal discourses (as it never could anyway, of course).

It is enlarged because its role is not 'only' to provide empirical data about specific ecological patterns and processes and predictions about their likely dynamics in the future, but should also play a theoretical–heuristic role (Jax, 2003). The latter role has been neglected hitherto. By the theoretical–heuristic role of ecology I mean that it should contribute to identifying knowledge gaps and uncertainties with respect to specific goals; should contribute to a clarification of research questions; and scrutinise the consistency between concepts and the 'responses' of physical reality. The case study on the use of ecological concepts in the EA of the CBD (Section 5.3) demonstrates the need for such a heuristic role in order to aid the implementation of ecosystem management strategies.

For assessing ecosystem functioning in a conservation context, the task of empirical ecological science is thus manifold. First, it has to characterise the given ecological patterns and processes of a site. Second, ecologists must assess different concepts of ecosystems in terms of their fit to reality, as given by these ecological patterns and processes. The third task is to assess likely trends of the ecosystem variables and analyse how they relate to the selected reference states of ecosystem functioning. Finally, ecologists should assess which options for sustaining or regaining the desired ecosystem functioning exist. Ecological investigations can and must also demonstrate the trade-offs existing between different ideas of ecosystem functioning. Ecosystem management is not a panacea for biological conservation. Protecting ecosystem functioning with a focus on ecosystem services, for example, is often blind towards protecting specific species. Ecosystem management focusing on one set of ecosystem services (e.g. food) often precludes the provision of other services (e.g. recreation), or has consequences for ecosystems and people in other places.

5.4.2 A network of hybrid concepts

'Ecosystem functioning' is value-laden and relative to specific ideas and definitions of an ecosystem. It requires a reference state that is not given by nature. In consequence, we may decide to abandon these terms completely as being beyond the realm of science. We may, however, use their hybrid characteristic of referring both to facts *and* values as a chance and an opportunity. In clearly acknowledging and stating the value aspects and not trying to search for an 'objective' and 'true' definition of ecosystem functioning concepts, we open them to discussion. We can search for a limited number of useful and adequate definitions: useful for our social purposes (such as sustainable resource use, biodiversity conservation, or other aims) and adequate for the ecological patterns and processes to which they refer.

The concept of ecosystem functioning is not unique in this sense. There is an increasing number of hybrid concepts within ecology and conservation biology. These concepts form a network of concepts. This network puts ecosystem functioning in a still larger context and shows how relations between ecological science and society have changed during the last decades.

The network originates in its current form largely from the biodiversity discourse. It includes different ideas about ecosystem functioning, as described above (ecosystem integrity, resilience, etc.), but also 'biodiversity' itself, 'ecosystem services', and even 'sustainability'. All these are hybrid concepts in the above sense. Although the rise of 'ecosystem functioning' has been driven largely by the biodiversity discourse, concepts of ecosystem functioning are increasingly beginning to gain a momentum of their own. The different concepts often reinforce each other. Particularly, the notions of ecosystem functioning and ecosystem services have given new urgency and political relevance to the importance of biodiversity. They are often also used to explain each other, as can be seen in many quotes throughout this book.

Characteristic for this network and the debates around its components is that it is pervaded or even driven by an oscillation between descriptive and experimental science on one side, and societal values on the other. The biodiversity debate is not just a continuation or revival of the old diversity–stability debate in ecology, but has been triggered by conservation (and thus societal) concerns (see Section 3.1). Nevertheless, it has initiated a lot of basic research on biodiversity and its relation to more specific ecosystem processes (as one meaning of 'function'). However, a new focus on both 'ecosystem services' and on the 'functioning' of the whole ecosystem has shifted the emphasis back to the more normative dimensions. With the introduction of concepts like 'ecosystem services' or 'resilience', environmental research and environmental policy witnessed a change in direction. 'Ecosystem services' highlights the contribution of nature to broader societal needs and economic purposes in a variety of applications, and therefore to human wellbeing. In particular, the Millennium Ecosystem Assessment (2003, 2005; see also Section 4.1) emphasises those characteristics of ecosystems that contribute to the benefits of the environment for the economy, or have economic and social consequences when degraded (production of food, waste treatment, flood protection, etc.). 'Ecosystem functioning', 'ecosystem services', and 'resilience' may be 'rising stars' of a new debate, necessary additions to the ongoing biodiversity discourse (e.g. to further demonstrate and legitimise biodiversity research), or just 'passengers' of the biodiversity debate. They may, on the other hand, even replace biodiversity as the key concept of environmental science and environmental policy.

6 · *Assessing ecosystem functioning*
Some existing approaches

No general agreement exists about to how to conceptualise and assess ecosystem functioning. There are a couple of approaches which attempt to measure some sort of ecosystem functioning, though under different labels. In this chapter I will investigate the most prominent of these approaches, particularly the concepts of ecosystem integrity, ecosystem health, ecosystem resilience, and ecosystem stability. To analyse the different approaches I will use the insights and analytic tools developed in the previous chapters. Special emphasis will be given to the question of how and under what circumstances the concepts can be applied to assess ecosystem functioning in the practice of conservation and resource management. A broader, more generally applicable conceptual framework for describing, classifying, and assessing different ideas and measures of ecosystem functioning will then be presented in Chapter 7. First, however, let me briefly summarise the results of the previous chapters to show our departure point.

6.1 Ecosystem functioning: the baseline

I have shown that ecosystem functioning means different things to different people. The basic types of understanding of ecosystem functioning are: (1) ecosystem functioning simply denotes some processes (or even properties) at the *level* of an ecosystem; (2) ecosystem functioning pertains to the performance of the whole system. Meaning 1 may sometimes serve to indicate Meaning 2, i.e. constitute a kind of proxy for Meaning 2; but, although often implicit, this is not always the case.

Even from the generic definition of an ecosystem (the ecosystem as an assemblage of organisms together with their abiotic environment) it is evident that the basic processes of any ecosystem are those performed by organisms (interactions with each other and with their abiotic environment). These are the driving forces of *any* kind of ecosystem functioning. While these are the *mechanisms* driving ecosystem functioning, another question is how to *assess* functioning in terms of the overall performance of an ecosystem. This may be (and is) often done by criteria other than the processes themselves – namely by the continuity of specific system states and products which are the outcome of the mentioned processes. These two aspects must be distinguished.

Which processes and properties count as pertaining to ecosystem functioning, however, differs, depending on the definition of an ecosystem embraced and on the background assumptions about the critical processes steering the performance of an ecosystem. Thus, Naeem (1998) describes ecosystem functioning as referring 'to the biogeochemical activities of an ecosystem or the flow of materials (nutrients, water, atmospheric gases) and processing of energy' (Naeem, 1998, p. 39), whereas Moulton (1999) states:

In certain uses the term 'ecosystem functioning' is restricted to the flow of energy and nutrients of the ecosystem in question. It is an obvious extension, however, to consider the details of the biotic interactions and the processes that alter and regulate the flow of energy and nutrients, such as alternative states of vegetation at the primary producer level and strongly interacting 'ecosystem engineering' [. . .] or 'key-species' [. . .] animals at the higher trophic levels. *(Moulton, 1999, p. 575)*

Both quotes relate to the processes and mechanisms, the 'collective activities' of an ecosystem and not as much to *assessing* ecosystem functioning. Besides just describing these collective activities, assessments of ecosystem functioning either deal with (1) different degrees of functioning; (2) different modes of functioning (in particular different 'alternative states' of ecosystems); or (3) the question of whether an ecosystem is functioning at all or if its 'dysfunctional', 'collapsed', or 'destroyed' – in terms of a dichotomy. All these assessments require *reference states*.

One may argue that an alternative state of an ecosystem is in fact the collapse of the 'old' ecosystem, which turns into another, new one. In the same way, different phases of a successional process may be considered as either constituting different ecosystems or as being only different expressions of the same ecosystem (this is also an old discussion in phytogeography, there with respect to plant communities). This is very much a matter of the perception of the specific observer and the specific background theories applied.

The idea of ecosystem functioning may thus denote *any* kind of processes, and in this manner any system that can be called an ecosystem is thus also a 'functioning' ecosystem. In most cases, however, the word 'functioning' refers to '*proper* functioning', that is, to a particular reference state, reference trajectory, or reference dynamic of the system. This is also implied in expressions such as 'safeguarding ecosystem functioning' or 'restoring ecosystem functioning'. To speak of a 'functioning ecosystem' (using 'functioning' as an adjective applied to ecosystems) would be almost tautological without a reference condition, without the meaning of *proper* functioning. The notion of the 'proper' functioning of ecosystems is the topic of this book and is paramount to many applied fields of the environmental sciences, such as ecosystem management or restoration ecology.

Defining and assessing ecosystem functioning in this sense of the term requires us to clearly state what is meant by an ecosystem and what is considered as a reference state. There are no convincing arguments that ecosystems can be defined as 'natural kinds' in the manner of organisms (or species) without involving problematic notions of teleology and ontology. Thus, the definitions of ecosystems and their reference states are also always a matter of *value decisions* (at least methodological value decisions). These decisions are not a matter for science alone – they are to a large degree shaped by societal choices. However, this does not render 'ecosystem' and 'ecosystem functioning' as purely arbitrary concepts or mere 'societal constructs'. Ecosystems are neither given by nature as such nor are they pure figments of human minds. They are – as described in detail in Chapter 5 – something that results from an interaction of physical reality and human conceptualisation.

In the following I will describe existing approaches for assessing the performance of whole ecosystems (i.e. ecosystem functioning) and scrutinise them with respect to the following questions:

(1a) Which definition(s) of an ecosystem is (are) the basis for the approach?

(1b) What range of possible other definitions are covered by the approach and (how) does it take account of the existing diversity of ecosystem definitions?

(2) What is considered as the reference state, dynamic or trajectory of the system and by which variables can it be assessed?

(3) How are societal choices included in defining the ecosystem and its functioning, and in assessing it?

The philosophical caveats described in the previous chapters (teleology, conceptual consistency, etc.) will serve as corrective tools for discussing the usefulness of the different approaches.

6.2 Existing approaches for assessing ecosystem functioning: ecosystem integrity, ecosystem health, ecosystem stability, and ecosystem resilience

A look at the biodiversity–ecosystem functioning (BEF) debate – being the discourse where the concept of ecosystem functioning is used most frequently today – shows that there is a great variety of variables used to indicate ecosystem functioning (Chapter 3). Considerable confusion exists here, because different ecosystem concepts are implicit in the discourse, with the definition of an ecosystem often being taken for granted or coming close to the generic definition of an ecosystem. Reference states are rather unclear. What is measured and assessed is mostly the change in some specific 'ecosystem' variable in comparison to a control, or change in levels of biodiversity. These variables are frequently taken as proxies for the overall performance of the system. No general scheme for conceptualising and assessing ecosystem functioning is available. The same situation is given for other fields where the functioning of ecosystems is conceptualised, such as restoration ecology or ecosystem management.

Some more refined attempts to assess the overall performance (i.e. functioning) of ecosystems are given by the concepts 'ecosystem integrity', 'ecosystem health', and 'ecosystem resilience'. All of these concepts are contested, not the least by the allegation of being teleological, or subscribing to a naive realism. As discussed in Chapter 4, this is an allegation that any concept of ecosystem functioning has to face if it goes beyond describing some selected ecosystem processes and tries to assess the (proper) performance of the whole system. It remains to be seen to what degree the different concepts must be considered problematic in relation to these issues.

In the following I will first briefly characterise the different concepts and then analyse them with respect to the questions above.

6.2.1 Ecosystem integrity and health

The concepts of ecosystem health and ecosystem integrity are closely related. They were both developed as a response to perceived shortcomings in the traditional evaluation of ecosystem conditions, such as assessment of species lists only or only selected physical and chemical parameters. The overarching idea, which both concepts share, was to have concepts that transcend partial measures of ecosystem variables and instead provide a 'holistic' approach to ecosystems and their proper condition (or functioning) as a whole.

Both concepts flourished during the 1980 and 1990s. They later diminished somewhat in importance, or at least were (and are) dealt with under different names. However, 'integrity' and 'health' have made their way into environmental legislation, where they are part of several acts and guidelines. In particular, 'ecological integrity' is a goal inscribed in the Clean Water Act of the USA (since 1972) and the Canadian National Park Law (since 1988) (Karr, 1996). Likewise, the European Water Framework Directive (WFD) (European Community, 2000) is following similar ideas (see the case study below). Thus, these concepts are of high practical relevance. As a consequence, there has been much debate about how to formalise and implement them. It is almost needless to say that, as with many other concepts, especially those which lend themselves to an intuitive understanding of their meaning, there is a variety of definitions for each of them. Ecosystem health and ecosystem integrity are sometimes seen as synonymous, sometimes as clearly distinct (Lackey, 2001).

Ecosystem integrity (also termed ecological integrity) first appeared as an explicit concept in connection with the question of how to assess and manage aquatic ecosystems. While there were some earlier attempts to define the 'integrity of water' (see papers in Ballentine and Guarraia, 1977), the most prominent approach to defining and measuring ecosystem integrity was developed by James Karr (Karr and Dudley, 1981; Karr, 1991, 1993). For Karr and those in his tradition, integrity is a state which comes close to, or is identical to, a pristine state of an ecosystem – a state unaltered and 'unharmed' by humans. Central to assessing ecosystem integrity is thus the comparison of a specific existing ecosystem state with a (pristine) reference state of the same ecosystem type. In Karr's own words, the integrity of an ecosystem is defined as: 'the capability of supporting and maintaining a balanced, integrated, adaptive community of organisms, having a species composition, diversity, and functional organization comparable to that of natural habitat of the region' (Karr and Dudley, 1981, p. 56). And he adds: 'A system possessing integrity can withstand, and recover from, most perturbations imposed by natural environmental processes, as well as many major disruptions by man' (Karr and Dudley, 1981, p. 56).

The core of ecological integrity in this sense is *biological integrity*, for which Karr developed his Index of Biotic Integrity (IBI) (Karr, 1991). Combined with chemical and physical integrity, biotic (or biological) integrity forms 'ecological integrity' (Karr and Dudley, 1981; Karr, 1996).

Integrity is thus closely tied to the idea of a biotic community in harmony with a specific site on behalf of its evolutionary history: 'A biota with high integrity reflects natural evolutionary and biogeographic processes' (Angermeier and Karr, 1994, p. 692).

Originally, Karr used his idea mainly for assessing the integrity of fish communities. He soon applied and extended the index to all species groups and also saw it as being transferable to other ecosystem types, not just aquatic ones. An important further extension was elaborated by Loucks (2000), who adapted the approach for forests.

While ecosystem processes and ecosystem dynamics are of high importance to Karr's understanding of ecological integrity (see above), his IBI relies completely on describing ecosystem *elements* as indicators for ecological integrity. These refer to three groups of metrics (Karr, 1991): (1) 'species richness and composition metrics' (total number of species, number and identity of selected species groups with similar ecological attributes); (2) 'trophic composition metrics' (percentage of individuals from different trophic groups); and (3) a metric related to the population density and condition of all organisms in the sample (total individual number, percentage of hybrids and percentage of individuals with diseases, damages, or anomalies) − referring to all fish in the original scheme. These attributes were selected to cover the 'range of ecological levels from the individual through population, community, and ecosystem' (Karr, 1991, p. 71).

In contrast (or in addition) to attributes focusing on the elements of an ecosystem, Loucks described 'mean functional integrity' (MFI) as a property characterised by the natural range or mean condition of 'two or more standard ecosystem functions' (Loucks, 2000, p. 181), i.e. ecosystem processes. As 'standard ecosystem functions', Loucks names the following processes: primary production, secondary production, hydrologic pumping/evapotranspiration, biomass decomposition, and nutrient/ mineral cycling (Loucks, 2000, p. 179).

Alternative approaches to conceptualising ecological integrity have been brought forward by James Kay (1991) and Robert Ulanowicz (e.g. Ulanowicz, 2000). Both focus strongly on interactions and flows, Kay with reference to thermodynamics, Ulanowicz emphasising 'ascendancy', an ecosystem property deriving from the trophic web of a system, expressed in terms of measures from information theory.

However, these and other approaches to ecosystem integrity are much less prevalent than those based directly on Karr's concept. His expression

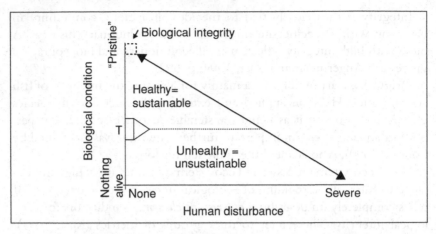

Fig. 6.1. Ecosystem integrity and ecosystem health as defined by Karr (2000). Karr characterises 'integrity' as the 'unharmed' or 'pristine' condition of a biological or ecological system. Based on societal values, the system may still be considered as 'healthy' when human influence increases and biological conditions deteriorate, even though it loses its integrity. Beyond a certain threshold of biological condition (T), the system ceases to be healthy, becoming 'unhealthy'. This threshold is reached when the biological condition is no longer 'sustainable', when 'neither the natural biota nor human activity can be sustained in that place' (Karr, 2000, p. 213). Figure reprinted from Karr (2000), p. 213, from *Ecological Integrity*, edited by David Pimentel, Laura Westra and Reed Noss. Copyright © 2000 by Island Press, reproduced by permission of Island Press, Washington, DC.

of the concept has surely been the most influential, including in terms of its use in environmental policy. In Europe, the WFD, adopted by the European Council and Parliament in 2000 as the major tool for protecting water and water-dependent ecosystems, builds on ideas largely similar to those developed by Karr (e.g. Moog and Chovanec, 2000) when defining the 'good status' of waters as the main environmental objective (see the case study below).

Distinctions between ecosystem integrity and ecosystem health are not clear-cut (Lackey, 2001). While some authors (e.g. Ulanowicz) shift between the two concepts, some seek to make strong distinctions. For Karr, for example, integrity is tied to being in the same state as the natural system. For him, integrity is a yes/no distinction. He therefore reserves ecosystem health – as a *gradual* measure – for all those systems that are influenced by humans, with healthy ecosystems being the goal of management. Ecosystem integrity is thus one end-point of a gradient of 'disturbance', the other being 'nothing alive' (see Fig. 6.1).

Others (e.g. Cairns, 1977; Regier, 1993; de Leo and Levin, 1997) deviate from this position in allowing also systems influenced and actively managed by humans to exhibit integrity. Regier (1993, p. 3) writes: 'A living system exhibits integrity if, when subjected to disturbance, it sustains an organizing, self-correcting capability to recover toward an end-state that is normal and "good" for that system. End-states other than the pristine or naturally whole may be taken to be "normal and good".'

De Leo and Levin (1997) even emphasise that '[m]easures of integrity must reflect the ability of ecosystems to maintain services of value to humans.'

For Ulanowicz (2000), ecosystem integrity likewise is a scalable, quantifiable – and thus gradual – property. Finally, Rapport even considered integrity to be a *prerequisite* for ecosystem health: '[T]he primary requirements for a healthy ecosystem are those of system integrity and sustainability' (Rapport, 1989, p. 128).

Ecosystem health is an even more ambiguous term than ecosystem integrity, as definitions of the former become quite elaborate and complex, but difficult to put into practice. Historically, some scholars attribute the idea of ecosystem health to Aldo Leopold (Mageau *et al.*, 1995), or even to the eighteenth-century physician and geologist James Hutton (see Calow, 1992). For sure, it has been present in the form of a vague metaphorical notion of the 'health' of the forest, waters, or even land(scapes) for a long time, taking the human (or at least mammalian) body as an analogy. However, in ecology it became a widely discussed concept only with the papers of David Rapport (Rapport *et al.*, 1985; Rapport, 1989). Subsequently, at least three journals were founded carrying 'ecosystem health' in their names (*Ecosystem Health*, 1995–2001; *Journal of Aquatic Ecosystem Health*, 1992–1996; *Aquatic Ecosystem Health & Management*, since 1989), of which only one persists to this day, with little explicit reference to 'ecosystem health' in its current papers. Rapport listed many possibilities for assessing ecosystem health (see below), but he avoided a formal definition. In 1992, for example, he wrote:

The short answer to this question is that a healthy ecosystem is whatever ecologists, environmentalists, and the public at large deem it to be [...] In the ecological realm one generally confers the connotation of 'health' to a state of nature (whether managed or pristine) that is characterized by systems integrity: that is, a healthy nature exhibits certain fundamental properties of self-organizing complex systems. *(Rapport, 1992, p. 145)*

There are, in principle, three major lines for defining ecosystem health. The first defines 'health' negatively, as the absence of 'disease'. It thus identifies different 'distress symptoms' (such as altered species compositions, pollutants, altered rates of productivity), building what has been called 'ecosystem distress syndrome' (Odum, 1985; Rapport et al., 1985). The second line follows more in the way of Karr – it defines health in relation to reference conditions for a 'normal' and 'good' state of ecosystems. The third approach was developed by Costanza and Ulanowicz (and was also used by the latter to define ecological integrity). It defines a healthy ecosystem by means of three components (Mageau et al., 1995), which the authors call 'vigour', 'organisation', and 'resilience'. 'Vigour' provides a measure for the overall 'metabolic activity' of the system (e.g. primary productivity), while 'organisation' depicts the complexity of interactions (measured e.g. by network analysis or Ulanowicz's 'ascendancy'). 'Resilience', following Holling, pertains to the 'ability to maintain its structure and pattern of behavior in the presence of stress' (Mageau et al., 1995, p. 204). Summarized by Costanza (1992, p. 248) in one sentence: 'To be healthy and sustainable, a system must maintain its metabolic activity level as well as its internal structure and organization (a diversity of processes effectively linked to one another) and must be resilient to outside stresses over a time and space frame relevant to that system.'

A definition trying to summarise and merge this type of definition and the first type into one common 'working definition' was given by Haskell et al. (1992, p. 9): 'An ecological system is healthy and free from "distress syndrome" if it is stable and sustainable – that is, if it is active and maintains its organization and autonomy over time and is resilient to stress.'

In the first issue of the journal *Ecosystem Health* in 1995, Rapport, although stating that the meaning of 'health' was still a matter for debate, provided an even broader approach, constituting a fourth line of defining ecosystem health. He listed seven (possible) 'key properties' of healthy ecosystems. These were: freedom from ecosystem distress syndrome, being resilient, being self-sustaining, not impairing adjacent systems, being free from risk factors, being economically viable, and sustaining healthy human communities (Rapport, 1995, p. 6).

However, all of these approaches can be put into practice with the help of very different variables. Also, they often refer to many concepts which in themselves are complex and ambiguous, such as 'stability', 'resilience', '(being) sustainable', and 'stress'. The meaning of all these concepts is discussed controversially in the literature.

In essence, 'ecosystem health', or the more specific expressions of 'forest health' (Warren, 2007), or 'river health' (Norris and Thoms, 1999) have remained rather vague and intuitive notions – and not clearly defined operational concepts. A concluding statement by Norris and Thoms, made in the overview article for a special issue on 'river health', seems typical to me, saying that '[i]t may not be necessary to define the term "river health" to gain scientific and management value from it' (Norris and Thoms, 1999, p. 205). References to ecosystem health are made time and again in the ecological literature, but without being narrowed down in some way. It seems instead that many different variables are tried out in order to give the concept some operational meaning. There is, however, no clear progress or maturation of the concept to be observed. What remains is a vague intuitive notion of what healthy ecosystems are: systems that are properly functioning, or in a 'good' (e.g. 'natural') state, which is then filled with different empirical variables that might confer some of this meaning.

So let us now approach 'ecosystem integrity' and 'ecosystem health' with the questions posed above. As there is large overlap between the two concepts, I will deal with them together here.

Definition of ecosystems
There are in general no clear definitions of what proponents of the concept mean by 'ecosystem'. We must at first assume that the ecosystem is defined in a rather generic way. Although there is frequently the assumption that a 'natural' ecosystem is self-sustaining, perpetuates itself without human aid through time, managed systems are also called ecosystems. Being self-sustaining is thus not a necessary part of most definitions. Usually, ecosystem-health concepts allude to a system model that is focused less on specific species patterns and more on processes and services useful to humans.

Reference conditions of the functioning (healthy, having integrity) system
The reference condition for ecosystem integrity in the sense of Karr is given by the region-specific natural ('pristine') condition of ecosystem types. Such 'typical' pristine ecosystems are assumed to exhibit integrity. Many other authors, and almost all those referring to ecosystem health, emphasise that the reference conditions depend on the specific social and ecological goals and settings. The stability (constancy) of different variables through time – in order for the system to be self-sustained or 'sustainable' – and the requirement of the system

to withstand disturbances are frequently parts of the desired reference conditions.

Not surprisingly, the breadth of possible measures and indicators for ecosystem integrity or health is considerable. Several authors argue that there is no silver bullet, no single universal indicator for assessing ecosystem health or integrity (de Leo and Levin, 1997; Boulton, 1999). No generally preferred set of indicators has emerged.

Normative dimensions and societal choices

The normative and vague idea of a proper state and/or functioning of an ecosystem clearly *preceded* the concept. Integrity and health, especially, are notions derived from the human sphere and have been used to convey a positive normative condition of an ecosystem. Laura Westra, particularly, has built a whole new environmental ethics on the basis of the 'principle of integrity' (Westra, 1994). Building at first on only intuitive ideas of ecosystem integrity, there were then efforts to provide meaning to the words. Westra, for example, states: 'At least some of the descriptions of ecosystem integrity (or synonymous ecological integrity) confer the impression that there is one (and only one) state of integrity of an ecosystem ('given by nature') and that different approaches to the concept are just different *measurement* approaches to this one integrity' (Westra *et al.*, 2000).

This impression is something that has often been criticised. Such criticism amounts to the allegation that ecosystem health and ecosystem integrity concepts promote a teleological view of ecosystems: that there is an optimum condition for ecosystems given by nature, implying a normative, 'good' state (Wicklum and Davies, 1995), even that ecosystems were thus depicted as a kind of 'superorganism' (Suter, 1993). In consequence, such a view of ecosystem health or integrity may hide the social character of environmental management goals by conferring the impression that these goals can and must be found in nature, and are not a matter of societal choice. This may be misused to argue for a *particular* condition of a 'healthy' ecosystem under the guise of 'objective', impartial science (Lackey, 2001). The term 'health' is especially laden with so much positive moral value in everyday language that it is difficult to argue politically against the goal of 'healthy ecosystems'. Who would argue in favour of 'illness'? This kind of use impedes the necessary discourse about goals and reference conditions of environmental management, hiding possible policy alternatives and possible trade-offs (what exactly is protected, what is not?) that exist between different alternatives of perceived ecosystem functioning.

In fact many, if not most, authors clearly state that ecosystem health and integrity are a matter of societal values (see quotes above). Also, James Karr, who promotes the most 'naturalistic' reference state for ecosystem integrity, clearly expresses that biological integrity is a 'societal goal' (Karr, 1996, p. 100), and not something we must follow as such. Nevertheless, the degree to which societal choices are seen as part of the process of both conceptualising and implementing ecosystem integrity and health varies considerably. If reference states (as in Karr's concept) are given by the pristine state of nature (with the assumption that these are properly functioning ecosystems), societal choices are, in principle, mostly restricted to the decision of whether to accept this as the basic reference condition for management. Somewhat more choices may also be made in terms of the method of implementing the assessment (and, where necessary, the achievement) of the functioning of ecosystems. In this case, most of the work of conceptualising reference states and their indicators is a matter for scientists. Other concepts of integrity or health, however, are wide open to stakeholder participation in formulating what should be defined as a functioning ecosystem (e.g. Rapport, 1992; Kay, 1993; de Leo and Levin, 1997; Meyer, 1997).

6.2.2 Case study: assessing 'good ecological status' in the European Water Framework Directive

The Danube is one of Europe's largest rivers, with a length of 2870 km. Its river basin encompasses more than 800 000 km^2, and is shared by 19 countries, most of them being members of the European Union. The quality of the water in such a river and that of its ecosystems should not just be managed individually by each country, but should be coordinated across national boundaries in an integrated approach. The drainage basins of many other European rivers likewise are shared by several countries. Therefore, the European Union has taken a river basin management approach in the WFD (European Community, 2000), which, since its adoption in 2000, has formed the backbone of European water legislation. The objective of the WFD is to reach 'good status' in all surface waters and groundwater in its member states, sometimes, as in the case of the Danube river basin, even involving states which are not – or not yet – members of the European Union, such as Croatia or Moldova. The WFD is a good example of how concepts of ecosystem functioning become relevant to the practice of conservation and natural resource management. More specifically, it is also an example of the application of *ecosystem integrity* as one approach for assessing ecosystem functioning. In this case study, I will first describe the WFD approach and how ecosystem functioning is assessed in this case. I will then discuss how – in the eyes of some critics of WFD practice – it might be assessed otherwise. This will serve to illustrate the strengths and weaknesses of the ecosystem integrity approach to ecosystem functioning.

As discussed in Chapter 5 in connection with the Ecosystem Approach (EA) of the CBD and also in the Yellowstone example (Chapter 4), (environmental) laws and regulations today contain many expressions relating to ecological concepts. Even more than in purely scientific texts, the terms and concepts used are not always clear in their precise meaning. If these texts are part of a legally binding and enforceable law that is to be implemented in administrative practice, the terms must be clarified. The WFD is such a text. Since its adoption in 2000 its implementation has become one of the foremost tasks of most European water managers, and has also inspired the work of many scientists. Scientific expertise has already decisively shaped the directive. The idea of maintaining (or, where necessary, restoring) ecosystem functioning ('ecosystem structure and function') plays an important role within the directive.

A brief outline of the WFD approach and its terminology
The WFD is a very demanding piece of legislation for the management of European waters. It differs substantially from the previous directives of the European Union aimed at safeguarding water quality. The older guidelines focused on different, specific uses of water, such as drinking water, water for irrigation, or bathing water. They defined threshold values for a number of chemical substances and for bacteriological criteria. The respective permissible values differ according to the specific kinds of uses. In contrast, the WFD aims at preserving and, where necessary, restoring an overall good water quality ('good status' in the terminology of the WFD) for all waters, irrespective of specific uses. Water quality for surface waters ('ecological status') is not defined just in terms of specific chemicals found in water, but is assessed using biological, chemical, and hydromorphological 'quality elements', thus focusing on the waters as ecological systems (ecosystems in the generic sense). This can also be seen in the description of the purpose of the WFD (Article 1). According to this article, the framework 'protects and enhances the status of aquatic

ecosystems', 'promotes sustainable water use', and 'aims at the protection and improvement of the aquatic environment' (in this order). Nõges *et al.* (2009, p. 203) also point out that the 'WFD represents a transition from chemically to ecologically based assessment of water quality. In principle, the ecological quality assessment must be based on the structure and functioning of aquatic ecosystems.'

By 2015, all water bodies in the European Union must be in what the directive defines as 'good status'. Limited exceptions are possible provided certain criteria are met. A monitoring programme must also be set up. The directive refers both to surface waters and to groundwater. In the following I will largely restrict myself to surface waters.

Defining 'good ecological status' in the WFD scheme
The idea of what constitutes the good status of a water body, and even more the so-called 'high status', which serves as a reference state, is in the tradition of James Karr's concept of ecological integrity and was explicitly inspired by it (Nõges *et al.*, 2009; Veronika Koller-Kreimel, personal communication, September 2009; see also Chovanec *et al.*, 2000). The status of a water body is described by its 'ecological status' and its 'chemical status'. Ecological status is classified as either high, good, moderate, poor, or bad. Chemical status is classified only as either good or bad.

The 'ecological status' of a water body is defined in general as:

'Ecological status' is an expression of the quality of the structure and functioning of aquatic ecosystems associated with surface waters, classified in accordance with Annex V. *(WFD, Article 2.21)*

To arrive at specific reference states and to allow the assessment of individual water bodies with respect to them, the WFD uses a *type-specific* approach. This involves a detailed procedure specified in Annex II of the WFD and in a specific Guidance Document that was written for this purpose (see European Commission, 2005). Guidance Documents are not legally binding, but constitute a common understanding of the interpretation of the WFD agreed upon by the EU member states. First, a *typology* of the different water bodies has to be developed, based on a variety of geographical, physical, and chemical factors. Based on that, *reference conditions* must be set up for each specific type of water body and, if possible, a 'reference network for each surface water body type' (WFD, Annex II, 1.3), containing a number of sites with a 'high status'. The latter is defined as the near-natural ('pristine') state of an ecosystem type under 'undisturbed conditions' (WFD Annex V, 1.2). In the Guidance Document on typology, reference conditions, and classification of lakes and rivers (REFCOND Guidance, see European Commission, 2003), 'disturbance' is explicitly defined as 'interference with the *normal functioning* of the ecosystem' (p. 78; my emphasis), i.e. an ecosystem (here a water body) with a high ecological status is assumed to be a normally (i.e. properly) functioning ecosystem.

High, good, and moderate status have to be characterised by a number of type-specific *quality elements*, namely biological elements, physico-chemical elements, and hydromorphological elements. For the different groups of biological quality elements (e.g. phytoplankton, macrophytes, or fish) the reference state (high status) is characterised by a 'taxonomic composition [that] corresponds totally or nearly totally to undisturbed conditions' (WFD, Annex V), while 'good status' means that 'there are [only] slight changes in the composition and abundance of [the specific] taxa compared to the type-specific communities' (WFD, Annex V). Type-specific species compositions and abundances are thus, as in Karr's biological integrity, a decisive characteristic of the desired quality states. Also, the focus is not on the specific species as such, but they are seen as indicators for the proper structure and functioning of the aquatic ecosystems. In contrast to other EU directives, such as the Birds and the Habitat Directives, the WFD is not specifically aimed at biological conservation.

Many different schemes for providing reference conditions have been proposed. They partly build on refinements and extensions of existing schemes developed in the different EU member states; most of them, however, have been developed specifically in response to the WFD requirements. The approaches differ, not the least in the degree to which they use specific species lists as references for specific types of water bodies. Sometimes, a longer list of specific species is given (mostly with some species emphasised as well-suited indicators for a specific type of water body), sometimes only higher taxonomic groups (or ratios of groups, e.g. large versus small copepods) are used. Biomass or chlorophyll concentrations are also applied as indicators for the quality of some biological elements. Mostly, multimetric measures are used for determining ecological quality (Lyche Solheim, 2005; Hering *et al.*, 2006; Nõges *et al.*, 2009).

Chemical and physical conditions should likewise conform to the type-specific, i.e. natural, reference conditions, but they are, in a way, in the service of the type-specific organisms and the ecosystem. They must 'not reach levels outside the range established so as to *ensure the functioning of the type-specific ecosystem* and the achievement of the values specified above for the biological quality elements' (WFD, Annex V; my emphasis).

How should a 'functioning ecosystem' be understood in the WFD context?
As is clear from the WFD text, ecosystem functioning plays an important role within the directive. Many authors involved in the WFD implementation process emphasise this aspect (Moss, 2007, 2008; Solimini *et al.*, 2009; Nõges *et al.*, 2009). Solimini *et al.* (2009, p. 144), for example, infer that: '[T]he classification systems should reflect any meaningful change in the structure of the biological communities and of the overall ecosystem functioning when affected by anthropogenic pressures.' As a consequence, they aim at methodologies that allow for the 'operational assessment of ecosystem functioning' (Solimini *et al.*, 2009, p. 144).

In contrast to some other terms, 'ecosystem functioning' is not defined explicitly in the WFD text and remains elusive to many scientists (Heiskanen *et al.*, 2004, p. 174). What we can derive from the text and the comments in the Guidance Document is that normally functioning 'high-quality' ecosystems are assumed to be given when the ecosystems refer to the biotic and abiotic conditions of (near-) 'natural' reference ecosystems, which have not been 'disturbed' by human influences.

There has been, and still is, severe criticism of the idea of defining and assessing good ecological status (and thus ecosystem functioning) by means of ecosystem integrity, i.e. type-specific detailed reference conditions derived from 'undisturbed' reference sites. Doubts have been raised both for practical and theoretical reasons. The objections raised mirror different ideas of what constitutes a functioning ecosystem.

On the one hand, criticism has been that 'true' ecosystem functioning cannot be assessed by means of species lists and type-specific chemical, physical, and hydromorphological reference values as defined in the WFD. The British freshwater ecologist Brian Moss, in particular, argues that the WFD is based on what he considers to be an outdated and unrealistic idea of natural ecosystems. Moss criticises the directive as presuming that 'ecological quality is independently measurable' by species composition and abundance 'reflecting ecological status', and that a specified not substitutable set of components exists for each type of water body (Moss, 2007, p. 383). He instead proposes a different idea of 'pristine' and functioning ecosystems:

To modern ecologists, a pristine ecosystem is one that maintains (sustains) itself independently of management and to do this, evolutionary mechanisms have built in resilience and flexibility to cope with inevitable natural changes. There is no simple, single formula which defines the ecosystem at a particular spot, no single list of species or even families or larger taxa, indeed no single overall structure, though the major features of physiognomy will be preserved unless climate changes markedly. *(Moss, 2007, p. 383)*

Moss does not argue against the assumption that 'pristine', 'natural ecosystems' are truly functioning ecosystems. He even speaks of pristine ecosystems like the Amazonian floodplains as 'independently functioning (high quality) ecological systems' (Moss, 2007, pp. 383 f.). Moreover, he also states that pristine systems should be considered as the appropriate reference for the good status of aquatic ecosystems. But he is at odds with the WFD (although, as he emphasises, not with its 'spirit') in terms of what characterises a natural and properly functioning ecosystem, and by which attributes functioning and ecological quality should be assessed. Instead of the 'secondary features' which he perceives the quality elements prescribed by the WFD to be, he postulates measuring 'fundamental characteristics' characterising ecological quality, which, for him, are '(a) efficiency in recycling scarce materials (nutrient parsimony), (b) characteristic physical and food web structure that ensures this parsimony and maintenance of the intactness of the system as a whole, (c) connectivity with other systems that also maintains intactness, and (d) a large enough size to allow resilience to change' (Moss, 2008, p. 33; see also Moss *et al.*, 2003 for a system devised for assessing the ecological status of shallow lakes). The rationale for this different approach is a non-equilibrium view of ecosystems, emphasising the occurrence of alternative stable states, and the assumption that climate change will make any attempt to maintain or restore communities with specific species compositions futile. But, as in the Yellowstone example, there seems to be another idea of 'natural' and even of 'pristine' implied by Moss than that in the WFD texts.

Moss' idea of a functioning ecosystem is thus highly different to that of the WFD and of its foundation, the concept of ecosystem integrity. Although both argue for natural reference states, their ideas of an ecosystem, and more specifically of a functioning ecosystem, differ with respect to there being fewer selected elements and a coarser resolution of the biotic elements (to speak in the language of the SIC model introduced in Section 4.3: physiognomy and processes, not specific species). The discussion thus bears some resemblance to the Yellowstone controversy described in Chapter 4.

A second argument against the type-specific approach of 'good ecological status' aims in another direction. More recently, some authors have argued that, especially in times of global change, ecological quality is not or should not be characterised by naturalness and specific species occurrences, but by the *services* and benefits which waters provide to humans (Dufour and Piégay, 2009). An answer given by the proponents of the WFD approach is that (naturally) functioning ecosystems in the sense of Annex V of the WFD will, in general, provide a better and broader suite of ecosystem services to society than ecosystems *designed* only for selected services (Koller-Kreimel, personal communication, September 2009; a similar statement, although based on his different idea of a functioning ecosystem, is made by Moss (2007, p. 390)).

Proponents of the ecosystem services approach to defining good ecological status also argue from a position along the lines of the EA of the CBD and its postulate that 'the objectives of management of land, water and living resources are a matter of societal choice' (see Section 5.3). This would mean a more open determination of quality objectives for each water body (or at least water body types) than is the case by having a rather fixed idea of natural conditions (however difficult to determine). This is connected to a stronger 'functional' approach to ecosystems in contrast to a more 'compositional' one (in the terminology of Callicott *et al.* (1999)), which Steyaert and Ollivier (2007) see as represented by the WFD. In fact, involvement of the public is included as a requirement in the WFD process (Article 14), but only in the implementation of the directive, not in the process of formulating specific quality objectives. Nevertheless, the construction of the WFD and the process of determining reference conditions and classifying the status of water bodies includes highly normative, value-laden decisions regarding the expression of an ecosystem that is considered as good and desirable – how nature *should* be (e.g. the idea that 'good' nature is nature without human influence; see also Steyaert and Ollivier, 2007).

In conclusion, the notion of ecosystem integrity forms the basis for understanding what constitutes a functioning ecosystem in the WFD: a system that is able to sustain a type-specific composition of species. This notion of functioning is, however, still contested as various different understandings of what constitutes ecosystem functioning exist, be it on behalf of different assumptions about ecosystems, different ideas of what is 'natural', or different ideas about the purposes of protecting and restoring the aquatic environment. The current approach, much more than most of the alternatives that have been proposed, has the advantage of being very specific, bringing forth a lot of valuable data, and – with the need to intercalibrate these data – also a lot of important theoretical and methodological discussions. While it would be politically highly problematic to challenge the whole approach of the WFD, the intense scientific and societal discussion the implementation process has elicited may lead to some conceptual refinements of some of the ideas on ecological quality in the long run, hopefully with a more explicit idea of what ecosystem functioning is to mean here. The WFD, in spite of some shortcomings, is an impressive piece of legislation. It is, in my view, currently the best solution to opening up the broadest array of options for the sustainable use of European water bodies and the life therein.

6.2.3 Ecosystem stability and resilience

There is a long tradition of assessing the stability of communities and ecosystems as a kind of measure for their overall persistence and performance. This has been especially expressed in the context of the now classical diversity–stability debate that took place during the 1960s and 1970s (Goodman, 1975; Trepl, 1995; see Section 3.1). It turned out, however, that 'stability' was (and still is) more a kind of overarching meta-concept, comprising very different and more specific concepts such as persistence, resilience, constancy, elasticity, etc., each of which also has several different meanings (Orians, 1975; Grimm and Wissel, 1997; Ives and Carpenter, 2007). This plurality of meanings, and the confusion that accompanies it (Pimm, 1984), was one of the reasons the debate on the relations between diversity and stability faded away in the early 1980s. Its resurgence in the 1990s in the form of the BEF debate took place in a different setting (see Chapter 3) and largely avoided the general term 'stability' in favour of focusing on more specific and more clearly defined response variables (but see Ives and Carpenter, 2007). Beyond stability in its meaning as the *constancy* of some variables over time, which is basic to almost any idea of assessing ecosystem functioning, another stability-related concept has gained strongly in prominence during the last decades. This is the concept of *resilience*.

While first introduced as a rather simple idea and used in a restricted ecological context by Holling (1973), resilience has been extended in many ways (see below). In contrast to the concepts of ecosystem integrity and health, it is also backed up explicitly by a large body of specific theory (or theor*ies*) (Gunderson and Holling, 2002; Folke, 2006; Brand, 2009a). The literature about resilience, in particular that promoted by the scientists of the so-called Resilience Alliance (www.resalliance.org), has become overwhelming. It is only possible to give a brief glimpse of the basic structure and ideas of this approach here – in order to allow a critical view of the concept – with respect to applying it as a possible measure of ecosystem functioning. This section owes much to the excellent dissertation of Fridolin Brand (2009a), who has provided what is currently the most extensive review and analysis of this field.

In his seminal paper of 1973, Holling introduced the resilience concept by pointing at the highly dynamic nature of ecological systems, where change is omnipresent and constancy seems to be the exception: 'But if we are dealing with a system profoundly affected by changes external to it, and continually confronted by the unexpected, the constancy of its behavior becomes less important than the persistence of the relationships. Attention shifts, therefore, to the qualitative and to questions of existence or not' (Holling, 1973, p. 1).

This means that the focus in these cases should be on the overall performance of the whole system instead of the quantitative constancy of specific patterns within it. Holling distinguished 'resilience' from 'stability'. He defined *stability* (very specifically) as 'the ability of a system to return to an equilibrium state after a temporary disturbance' (Holling, 1973, p. 17), emphasising the speed of return to this equilibrium point as a measure of stability (in this sense of the term). In contrast, he saw *resilience* as a 'measure of the ability of these systems to absorb changes of state variables, driving variables and parameters and still persist' as expressed by the persistence of relationships within the system (Holling, 1973, p. 17). Resilience in this way does not assume an equilibrium condition of a system. In a later publication Holling used the term 'resilience' for both of these concepts, qualifying the first as 'engineering resilience' and the second as 'ecological resilience' or (still) simply 'resilience' in the strict sense (Holling, 1996). In the discussion about resilience, it is the latter meaning that is predominant, as the theory into which the concept is embedded today does not presume equilibrium systems. When talking about resilience here, I will refer specifically to *ecological resilience* when using the term 'resilience'.

But Holling's concept of ecological resilience has been modified, redefined, and extended in several ways (see Brand and Jax, 2007 for an overview of major definitions). Extensions refer to (1) inclusion of a broader system definition, thus dealing not only with ecological systems but also with social systems and coupled 'social–ecological systems'; (2) inclusion of normative aspects in varying degrees; and (3) use of resilience both as a specific measure and as a kind of metaphor (or perspective) applied to different kinds of systems (Brand and Jax, 2007).

There are substantial doubts over if and under what circumstances or conceptual modifications resilience may be transferred to social and social–ecological systems. Of course almost all ecological systems on this planet today are influenced by humans in some way, and we certainly have to look for methodologies for linking to human decisions and behaviour when dealing with ecological systems. This is even at the core of this book. There should, however, be caution when it comes to investigating and modelling human societies *with methods and theories derived from ecological systems*. Social systems are not deterministic systems, but are subject to many factors not present in ecological systems, such as complex decision processes, involving negotiation processes and compromises, and power relations, but also highly irrational behaviour (see Hornborg, 2009). It has been a problem for many systems theories that they started with a very restricted domain (e.g. Bertalanffy describing the individual organism, H. T. Odum describing the world strontium cycle or energy transfer in an aquatic ecosystem) and later ended up in systems theories that were extended and applied to almost everything (as Bertalanffy's General Systems Theory of 1968, or Odum's energy-related theory flow models, claiming to explain even wars and religion, as in Odum (1971)). Resilience theory has likewise been seen as applicable to questions as different as explaining our inability to stop the 'biological decline of the Everglades', the worldwide threat of bacterial antibiotic resistance, or the collapse of the Roman Empire (Walker and Salt, 2006, p. 117).

In order to avoid this discussion and to keep the approach in line with the other ideas about ecosystem functioning discussed here, I will restrict my considerations largely to resilience as applied to *ecological systems*, excluding not the *impacts* of humans, but excluding society and its processes as a part of the system whose functioning has to be assessed.

Even among the more ecological definitions of resilience, there are different ideas as to what resilience means and consequently how to assess or even measure it. Thus Walker *et al.* (2002) state: 'Resilience, therefore, is the potential of a system to remain in a particular configuration and to maintain its feedbacks and functions, and involves the ability of the system to reorganize following disturbance-driven change.' Walker and colleagues then name the following as defining characteristics of a resilient system: the capacity to absorb disturbances, the capacity for self-organisation, and the capacity for learning and adaptation. From the wording it does not become completely clear as to whether the latter 'defining characteristics' are really meant as properties that have to be fulfilled to call a system 'resilient', or if these are just attributes that are (always or commonly) found as additional (resilience-conducive) properties of a system, the resilience of which is defined by other means. This is, however, crucial when it comes to *assessing* resilience.

In any case, and for all definitions, (ecological) resilience is not just a measure of ecosystem functioning under constant or 'undisturbed' conditions, but always refers to the ability to deal with external changes or disturbances. Thus, as Carpenter *et al.* (2001) point out, in order to measure ecological resilience, it is necessary to first clarify to what variable(s) resilience refers, and to what kind of disturbances ('resilience of what to what?'). Further developing and formalising this postulate, Brand proposed a three-step procedure for measuring ecological resilience. These steps are (Brand and Jax, 2007; Brand, 2009a): Step 1 – characterise what exactly should be resilient, i.e. which system is considered and which variables of the system (e.g. specific services it provides) should be preserved; Step 2 – characterise against what kind of disturbance(s) or disturbance regime(s) the system should be resilient; Step 3 – characterise the variables by which resilience should be measured.

All of these steps are far from trivial when it comes to putting them into practice. Step 1 includes value decisions: sorting out the variables of interest. This might sometimes simply refer to methodological values, but when, as is common, ecosystem services are included, it also refers to morally relevant societal values (see Section 5.2 for this distinction). Step 2 raises questions about how to quantify disturbance regimes (White and Pickett, 1985; White and Jentsch, 2001), not the least as to whether resilience in the specific case should be defined as 'general', i.e. referring to all kinds of possible disturbances, or 'specific/targeted', i.e. restricted to a specific type of disturbance (e.g. disturbance by fire, but not by wind) (Walker and Salt, 2006, pp. 120 ff.). The latter will, in general, be more appropriate and feasible.

An important difference between resilience and most other ways of assessing ecosystem functioning is that it relates not just to constancy or trends but to *capacities* of the system (to absorb disturbances, to reorganise, etc.). This is much more ambitious than, say, measuring the degree of similarity to a reference state, as in most concepts of ecosystem integrity. If it was possible to really assess these capacities, this measure would also possess a much higher predictive potential than other measures of ecosystem functioning. However, as most authors agree, ecological resilience cannot be measured directly, only by means of surrogates (Carpenter *et al.*, 2005; Bennett *et al.*, 2005). Several different surrogates have been proposed and there is yet (if there ever will be) no silver bullet for the task. Brand (2009a) provides a review of the most relevant and widely discussed resilience surrogates and (indirect) measurement approaches. The idea of determining a system's identity (not exactly the same as the SIC concept as introduced in Section 4.3) is rather new in resilience studies (Cumming and Collier, 2005; Cumming *et al.*, 2005) and has not been implemented much. The authors propose first clarifying the elements of a system that are considered essential to identity, then assessing the potential for changes in identity in a given context of likely environmental dynamics, especially disturbances, using a scenario-based approach of possible futures. The surrogate for systems resilience is then given as the 'likelihood of a change in identity (and its magnitude)' (Cumming *et al.*, 2005, p. 984).

The two other major approaches towards assessing the resilience of a system refer to: (1) estimating the distance of the system from a *threshold*, where it collapses and/or turns into another system; or (2) measuring that amount of *resilience-conducive mechanisms and properties* in the system.

The basic assumption of the threshold approach is that the system is highly dynamic and thus exists within a broad range of possible system states, which nevertheless are within a common 'domain of attraction' or 'basin of attraction', characterised by the same kinds of interactions between the system elements (see Fig. 6.2). Resilience, then, is the capacity of the system to stay within the domain of attraction in the face of disturbances, even though the parts of the system may change (e.g. in terms of population sizes or even species). Once a particular threshold is crossed, however, the system either ceases to exist or moves into another domain of attraction, i.e. an alternative stable state (see the case study in Section 5.1; Holling, 1973 also referred to this idea). The distance of the system to such a threshold is used as a measure for its resilience, described as the position of the variables constraining the system's behaviour within the domain of attraction. This approach is, however, hinged on many difficult assumptions (such as the existence of thresholds and alternative stable states of a system) and far from easy to operationalise (e.g. Bennett *et al.*, 2005; Brand, 2009b).

A third way to assess resilience has been proposed by Brand (2009a) and foreshadowed in Carpenter and Cottingham (1997) and Nyström

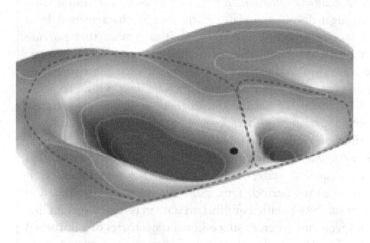

Fig. 6.2. A resilience landscape. See text. Source: Walker *et al.* (2004), with permission from *Ecology and Society* and Brian Walker.

(2006). This approach determines resilience by the amount of resilience-conducive properties present within a system. Properties that have been described as such are, for example, population diversity, the maintenance of species and life-history traits important for specific ecosystem processes, the amount of ecological redundancy, or the occurrence of intermediate disturbances at small scales (Brand, 2009a). Many of these properties further the 'potential for functional compensation' (Brand, 2009a, p. 174), which in turn is seen as decisive for ecological resilience.

None of the three approaches towards assessing ecological resilience is really established or at a mature stage. All of them face high uncertainties and still require a considerable amount of empirical and theoretical research.

Box 6.1 *Resilience terminology applied: the resilience of a coral reef*

The concept of ecological resilience is a rather complicated one. The most important aspects can be explained best by referring to a specific empirical object. Coral reefs are one object to which the resilience concept is frequently applied in the literature (e.g. Nyström *et al.*, 2000, 2008; Mumby *et al.*, 2007).

(*cont.*)

Coral reefs have been observed to exist in two or more *alternative stable states (domains of attraction, regimes)*, such as a coral-dominated state and an algal-dominated state. Each state is characterised by a number of species with specific roles, such as species that provide physical structure, different types of producers, types of herbivores, etc. There is no slow and gradual transition between the two stages, but a sudden *phase shift* (often also called *regime shift*). Coral reefs are subjected to a variety of natural disturbances at many levels, ranging from patchy grazing to hurricanes. The long-term temporal and spatial distribution of the individual short-term disturbance events is the *disturbance regime* to which the reef is subjected. In spite of such disturbances, a specific state, such as the coral-dominated state, is maintained over a long period. This means the system is considered to be highly *resilient*. No specific equilibrium state in terms of species distributions is given and different successional trajectories of a disturbed patch are possible. But local extinctions of some species are compensated for by, for example, rapid larval dispersal or by replacement with species with similar roles. For example: herbivorous species impede algal growth and facilitate recruitment of coral larvae. High diversity of herbivorous species allows rapid replacement of single species from this functional group when these are affected specifically by some disturbance. This is an example of an important *resilience-conducive property* of the system. Specific spatial configurations of patchy species distributions and water currents are likewise important for promotion of resilience. Species composition and abundance will change, but the overall interaction between particular functional types of species will be maintained. Some types of disturbance, however, such as by pathogens, hurricanes, or strong overfishing, can reduce the resilience of the system and drive it beyond a *threshold*. The threshold may be characterised (or indicated) by a certain amount of algal cover or coral cover. If, for example, herbivores are not able to inhibit the growth of algae any more, algae can become so abundant that even after the return of herbivores, they cannot be controlled by them any more. Algae can even overgrow and kill existing coral colonies; they also inhibit new coral larvae recruitment. The threshold has been crossed, after which the system is in a new stable state (domain of attraction), dominated by algae and other specific associated types of organism.

The fact that a system is considered resilient (or, in the terms of the Resilience Alliance, remains within a basin of attraction) might be considered as being synonymous with the statement that the ecosystem is (properly) functioning. Resilience might, however, be seen only as *one of several* characteristics of a functioning ecosystem, as in some definitions of ecosystem health (e.g. river health (Boulton, 1999)) or ecosystem integrity.

As has become clear, resilience is embedded into a complicated and highly contested theoretical framework. Looking at this framework does not make the assessment of ecosystem functioning easier, but actually complicates the task. This refers especially to the notion of 'complex adaptive systems' in which ecosystems (or social or social–ecological systems) are seen as characterised by a complex 'adaptive cycle', within which they change between different systems states and also experience cyclic phases of 'creative destruction' (a term first known from innovation theory within economics) and reorganisation. Here resilience waxes and wanes, but one could still consider the whole cycle as what makes/characterises a functioning ecosystem. This requires, of course, a highly extended temporal (and often also spatial) scale of observation. It surely also renders assessing whether an ecosystem is functioning much more difficult. It presupposes, e.g. that ecosystems really have alternative stable states, which is far from evident and general. And also, how do we distinguish if a collapse of a system is part of its proper functioning (part of a functioning adaptive cycle) or a sign of 'malfunctioning' – without waiting to see its actual further development? The situation is further complicated, as such adaptive cycles might be perceived as part of a nested control hierarchy, labelled 'panarchy' by its proponents (Gunderson and Holling, 2002, especially Chapter 3). That is, functioning and resilience may also be perceived on different scales. I will come back to the issue of scale as a matter of general importance for the idea of ecosystem functioning (Section 7.4).

So let us now take a look at how resilience theory relates to ecosystem functioning in terms of the definition of the ecosystem and its reference states, and in how it deals with normative issues.

Definition of ecosystems
The definition of the ecosystem within resilience research is not really clear. Quite typically, a review paper by Elmqvist and colleagues (among them some of the most prominent resilience scholars) starts with the statement: 'Ecosystems are complex, adaptive systems characterized by

historical dependency, non-linear dynamics, and multiple basins of attraction' (Elmqvist et al., 2003, p. 488). We might ask if that is meant as a definition of an ecosystem, which would, however, necessitate (at least) specifying what the terms (especially 'adaptive system') mean. In any case, the definition is not sufficient for delineating an ecosystem in space. While the above statement may surely not be an exhaustive definition of what the authors mean by an ecosystem, it remains nevertheless necessary to state more clearly what they mean by it, because the statement is surely not valid for all systems that ecologists call 'ecosystems' or for the generic definition given in Chapter 4. So we are faced time and again with a vague idea of what the ecosystem is and to what kinds of systems resilience statements apply. Is resilience a useful measure for all ecosystems (in the generic sense) or only for those that are systems that fulfil the criteria in the above quote?

All in all, ecosystems appear to be seen as 'given' by nature and not in need of being defined in more detail (see also Kirchhoff et al., 2010). What we can say about the idea of an ecosystem behind most resilience studies is that the ecosystem is considered as something highly dynamic, which cannot be easily pinpointed by very specific patterns (e.g. of species) or processes. When we look at the most-cited examples for resilient systems, especially those described as exhibiting alternative stable states, the physiognomic character of the whole system (in contrast to species identities) appears to be one decisive aspect for defining a system in terms of its basin of attraction.

The idea of nested hierarchical systems ('panarchy') mentioned above makes it even more difficult to see what the actual system is, i.e. what the system is that is to be perpetuated in some way, that is to be considered as either resilient or not resilient.

Like many other concepts, such as the ecosystem, ecosystem health, disturbance, and many others, 'resilience' is also often used more as a metaphor or a guiding principle (perspective) applied to management, instead of as a measurable property of ecosystems (see Box 5.1). This makes it even more complicated to assess the meaning(s) attached to the 'ecosystem' than it already is.

Reference conditions of the functioning (resilient) system
If the ecosystem in focus exhibits alternative stable states, reference conditions are given rather evidently by the thresholds of its domain of attraction, once these are defined or 'found'. Any changes of the system that do not cross the thresholds and change the system into another

state are negligible in terms of the resilience of the system (and thus of its proper functioning). Frequently, reference conditions are defined on the basis of ecosystem services provided by the system. As long as their benefits are delivered by the system (e.g. fish resources, wave protection, and recreation from coral reefs) it is still in the basin of attraction (Walker et al., 2002; see also Brand, 2009a). The question is, however, how to determine reference states for systems which do *not* exhibit regime shifts. It may even be questioned whether the concept of resilience is applicable to such systems at all.

Normative dimensions and societal choices
Resilience is often applied in a normative context or even in a normative manner (see Brand and Jax, 2007). But resilience is not *desirable* per se. There can be highly resilient states of ecosystems which are very undesirable from some human perspectives, such as algal-dominated coral reefs (are they still coral reefs at all then?) or hypertrophic lakes. Mostly, however, resilience *is* seen as a desirable property of a system.

A discussion of societal choices and participation takes place mostly in the context of adaptive management approaches (e.g. Folke et al., 2002). Adaptive management approaches are often seen as a consequence of resilience theory. There are also approaches towards stakeholder involvement at every stage of resilience management in social–ecological systems, including the definition of the system models themselves (e.g. Walker et al., 2002). Discussions about how resilient *ecosystems* should be conceptualised, however, taking into account different value decisions, seem to be largely absent in the literature on ecological resilience.

6.2.4 Other approaches

I have sketched out the most important approaches relating to assessment of ecosystem functioning. There are several others that have gained less attention. Some of these should at least be mentioned here.

A concept closely related to resilience is that of *system vulnerability* (Adger, 2006; Villagrán de Leon, 2006). The Intergovernmental Panel for Climate Change (IPCC) defined 'vulnerability' as: '[t]he degree to which a system is susceptible to, and unable to cope with, adverse effects of climate change, including climate variability and extremes' (IPCC, 2008, p. 89). As a technical, explicitly defined term the concept is mainly used to describe and assess social–ecological or social systems. It is often seen as being the inverse of resilience (Villagrán de Leon, 2006, pp. 50f.). This is also the meaning implied most frequently in ecological studies. In ecological contexts the term is, however, used mostly in a rather informal way, and is applied less to ecosystems, more to individuals and populations. Those applications within the ecological realm that provide explicit definitions of vulnerability are mostly related either to risk assessment, e.g. in ecotoxicology (e.g. Williams and Kapustka, 2000), environmental monitoring (Boughton et al., 1999), and/or conservation planning (Wilson et al., 2005). In detail, the approaches differ, but they have in common that, like resilience, they refer not only to the 'normal' performance of ecosystems, but to their behaviour in response to disturbance or stress.

There is another direction of ecological theory, which approaches measuring ecosystem functioning by means of *thermodynamic characteristics* of the systems. The idea of 'ascendancy' developed by Robert Ulanowicz has already been mentioned in connection with ecosystem integrity. Also on the basis of a thermodynamical perspective, and related to optimisation theory and self-organisation theory, the group around Sven Jørgensen has developed the concept of 'ecological orientors' (e.g. Müller and Jørgensen, 2000; Ludovisi et al., 2005), understood as 'aspects, notions, properties, or dimensions of systems which can be used as criteria to describe and evaluate a system's development state' (Müller and Jørgensen, 2000, p. 562). A large number of orientors have been named, starting from thermodynamic orientors via information and network theoretical ones through 'functional, system dynamical and organisational' orientors, by which 'a holistic picture of the ecosystem's state' can be derived (Müller and Jørgensen, 2000, p. 562). One of the most prominent single indicators for ecosystem functioning here is the 'exergy' of an ecosystem, describing 'the distance from [thermodynamic] equilibrium of an open system' or the 'maximum amount of work that can be extracted from an open system' (Libralato et al., 2006, pp. 572f.). It has been suggested and applied as a 'holistic' indicator for ecosystem states and/or performance (e.g. Jørgensen, 1992; Silow and In-Hye, 2004; Libralato et al., 2006). These thermodynamic approaches are often linked (explicitly) to concepts of ecological integrity and/or ecosystem health. Like ecological resilience, they view the ecosystem as highly dynamic (in contrast to static or equilibrium views) and include these dynamics in their concept of ecosystem functioning. Nevertheless, their reference state is frequently the 'natural' state – or in many cases the 'natural' trajectory/development – of ecosystems.

Occasionally, attempts have been made to introduce *ecosystem sustainability* as another concept for the overall (long-term) performance of ecosystems. Chapin *et al.* (1996, p. 1017) defined this term by characterising a 'sustainable ecosystem' as one that 'over the normal cycle of disturbance events, maintains its characteristic diversity of major functional groups, productivity, soil fertility, and rates of biogeochemical cycling'. The authors explicitly do not refer to the use-oriented dimensions of sustainability. However, economic and social criteria — beyond purely environmental sustainability — are at the core of most definitions of 'sustainability' today, being mostly debated in connection with 'sustainable development' (e.g. Parris and Kates, 2003). That is, sustainability research, while becoming a discipline of its own, is dealing not only with ecosystems, but also with human societies and their environment, and their relations with each other. The fact that 'sustainability' is mostly understood in this broader context may be one reason why 'ecosystem sustainability' has not become a technical term in ecology and conservation biology. Like 'ecosystem vulnerability' the expression is mostly used rather loosely to describe ideas of the long-term continued performance and/or use of ecosystems.

6.3 Conclusions from this chapter

Within the cluster of concepts that aim to describe ecosystem functioning as the overall performance of ecosystems, a variety of approaches exist. The different concepts (such as ecosystem integrity, health, and resilience) are often considered as connected or even part of each others' definitions, without, however, the existence of a clear structure of such connections. This is because each of the concepts appears in various definitions, each with different underlying assumptions, not the least about the ecosystem concept, about reference states, and about their normative status.

While all concepts of ecosystem functioning are by necessity formulating criteria of *constancy* for some ecosystem variables as decisive for the 'proper functioning' of the systems, the resilience concept goes one step further. It does not only consider the condition of the system at particular moments in time, but in addition involves the responses of a system to external disturbances ('resilient to *what?*'), expressed as the *capacity* of the system to maintain functioning in the face of these disturbances. As such, the concept is much more ambitious than other concepts relating to ecosystem functioning, but also, in consequence, much more difficult to apply to specific situations.

All of the concepts are highly normative, although the degree of normativity varies – each concept's idea of ecosystems and, especially, of reference states is subject to societal decisions.

Ecosystem functioning, resilience, integrity (and, even more, vulnerability) are often used in a very vague manner. This is not a problem as such, but authors often move back and forth between the broad and narrow meanings of the concepts without a clear distinction. Are we dealing with something that can be put into practice, can be measured? Or are we dealing with a general and vague idea? Either may be appropriate, but in different circumstances (see Box 5.1). Using, 'ecosystem health', 'resilience', and 'adaptive cycles' as general metaphors for thinking about the management of ecosystems is legitimate, but is not to be confused with measuring or predicting resilience in specific management situations, where action depends on empirical data assessed for operationalising the concept.

Taken one step further, 'resilience', 'ecosystem functioning', or 'ecosystem health' have even become *boundary objects*. Within the field of science and technology studies, this signifies a term that facilitates communication across disciplinary borders by creating a shared vocabulary, although the understanding of the parties would differ regarding the precise meaning of the term in question (Star and Griesemer, 1989). Boundary objects can coordinate different groups without a consensus about their aims and interests. If they are both open to interpretation and valuable for various scientific disciplines or social groups, boundary objects can be highly useful as a communication tool in order to bridge scientific disciplines and the gap between science and policy (Eser, 2002).

Indeed, it is this vagueness and malleability – i.e. the potential variety of interpretations or applications of the term – that makes boundary objects politically successful. For example, the boundary object 'biodiversity' has been very successful in bringing together people from highly divergent scientific fields and highly divergent interest groups (Eser, 2002; Brand and Jax, 2007).

But there is also a danger in this. Boundary objects can be a hindrance to scientific and political progress. They may become completely unclear if too many different meanings are used, or if only vague generic meanings are used – and thus useless for precise scientific tasks (such as assessing whether an ecosystem is considered to be properly functioning or not). On the political side, boundary objects may even hide conflicts and power relations when different individuals agree on the importance of protecting biodiversity or functioning ecosystems, when they actually mean different things by these terms.

Finally, different *terms* denoting concepts of ecosystem functioning are also different with respect to the implicit messages they convey, especially in the policy arena and to the broad public. Different words carry different connotations, they create different impressions and expectations, e.g. with respect to origins of the norms implied by them: taken from nature versus set by society. These connotations are largely independent of how scientists define the concepts in their papers. In this case, we are really dealing with a matter of semantics, with *words* and their secondary meanings in everyday language. For this reason, the world 'health', in particular, should be avoided in connection with ecological systems. Although it facilitates communication to non-scientists at first sight, by alluding to a 'normal', nature-given state, it actually impedes rather than facilitates a discourse about the societal aims of conservation and natural resource management.

7 · *Putting ecosystem functioning concepts into practice*
A classification and some guidelines

In this chapter I will draw together the discussions in this book in order to show a general procedure for putting the concept of ecosystem functioning into practice. For this purpose I will extend the discussion of existing concepts, such as ecosystem integrity or ecosystem resilience, to include a broader array of different ideas of what a functioning ecosystem is and how we can assess ecosystem functioning. Using the SIC scheme, which I introduced in Chapter 4, I will first discuss four common types of ecosystem concept and describe what 'proper functioning' means for them (Section 7.1). Following this classification I will then develop some guidelines on how to conceptualise and assess ecosystem functioning in conservation practice (Section 7.2), followed by a case study from restoration ecology (Section 7.3). Finally, I will draw some general conclusions on the potentials and limitations of the ecosystem functioning concept (Section 7.4).

7.1 Ecosystem functioning concepts in practice: a classification

Instead of starting with a specific index or a very general idea of ecosystems, we may go the other way and start with the definitions of (intact or functioning) ecosystems as they are commonly implied in conservation strategies and in scientific papers. These ideas often do not refer to the *mechanism* by which an ecosystem 'functions', but instead characterise a functioning system by its outcome, i.e. the persistence or continued existence of specific *properties* or *products* that are thought to characterise a functioning ecosystem – acknowledging that the system is dynamic internally. Even the most dynamic view of what an ecosystem is must presume that there is at least something of ecological significance that has to remain constant through time. Otherwise, we could not perceive of it as an object about which we can talk and whose properties we can

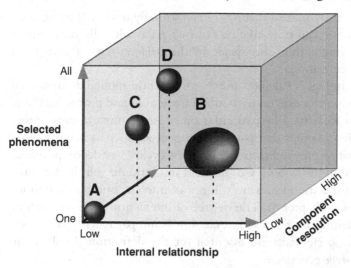

Fig. 7.1. Four common ideas of functioning ecosystems. A: the generic-type ecosystem; B: the process-focused-type ecosystem; C: the physiognomic-type ecosystem; and D: the species-specific-type ecosystem. See text.

measure. This is the idea of what we (Jax *et al.*, 1998) have called the 'self-identity' of an ecological unit.

Figure 7.1 summarises four different common ideas of what functioning ecosystems are and by which criteria they are defined – in particular in a conservation and natural resource management context. I will first briefly characterise these ideas and subsequently describe each of them in detail, based on the background of what was discussed in the previous chapters. I have given each of the four types a nickname, in order to speak about them in a way that may allow the reader to more easily remember them throughout the text. Please note, however, that these are just convenient labels which do not refer strictly to specific discourses (e.g. about wildness/wilderness) that are also carried out under these names.

The *generic idea of ecosystem functioning* derives from the generic definition of an ecosystem (Chapter 4). This definition is depicted by sphere A in Fig. 7.1. The ecosystem is preserved simply as a system of interacting biological objects and their environment, without much specification of the objects themselves, beyond that they are organisms. The kinds (i.e. species) of organisms are not of special importance here (i.e. low component resolution), and the degree of required internal relationships is similarly low (every kind of process will do). In the context of biological conservation, this kind of ecosystem may be useful for the management of wilderness areas, or areas that have been strongly impacted by humans but where 'nature can now take its course'. As a culturally determined subtype, one might thus also speak of the *wildness-type* of ecosystem functioning (see below).

Ellipsoid B in Fig. 7.1 depicts another frequently applied definition of ecosystems, which focuses on particular interactions and processes. Here, the ecosystem is defined by particular functional compartments, interacting in such a manner that particular processes (such as biomass production, predator–prey interactions, nutrient cycling, or decomposition) are maintained in the system. Component resolution is thus higher than in type A, but specific species are still not of interest, only specific functional types (see Chapter 3). The degree of interaction is higher than in many other definitions because interactions – and particular feedbacks – between specific elements are essential for the definition. I call it the *process-focused-type* ecosystem.

An important variety of this type can be labelled as the *ecosystem-services-type* ecosystem. Here the processes to be maintained over time are those that lead to the provision of ecosystem services – such as food production, clean waters, or erosion control – by the system. The provision of the desired services is then the indicator for the proper functioning of the system. This kind of definition is sufficient when the aim of ecosystem management is to provide benefits for humans.

Sphere C in Fig. 7.1 depicts a third type of ecosystem definition, demanding higher values in the three axes. For example, a beech forest ecosystem or a grassland and the essential interactions that perpetuate such systems are to be protected. The aim is to protect a large ecosystem that is 'typical' for the area, without needing to preserve all constituent species in the long run, except for some conspicuous and dominant taxa such as *Fagus* trees (beech forest ecosystem) or specific dominant grasses (grassland ecosystem). Particular types of taxa (indicator species, keystone species, or 'umbrella species' (see Simberloff, 1998)) are thus already part of the definition. This – physiognomic – view of ecosystems is perhaps the most common one in the practice of conservation and resource management, but also common in the everyday parlance of ecosystems and their functioning. I will thus speak of this model here as the *physiognomic-system type*.

Sphere D in Fig. 7.1 illustrates a concept of the ecosystem defined by all species occurring in an area. Interactions themselves are protected mostly for the sake of conserving the interacting components. These may be those species that are present in a protected area at date t (e.g. the date at which the measures start, or an important historical date) or – much more difficult to determine – all species that are considered as 'typical' for a particular site. The aim here is to perpetuate all species, without fixing particular growth rates or dwelling places, abundances, or specific ratios between species. This aim is formulated, for example, in some national parks and corresponds to the strategy of ecosystem management in Yellowstone National Park (see Section 4.3 and Jax, 2001). I will call this the *species-specific type* of ecosystem.

These four ideas of what constitutes a functioning ecosystem are not exhaustive but, in my view, are the most important ones; there may be intermediate types and completely different ones for specific purposes. The types described above are not different *empirical* types of ecosystem in the sense that we frequently distinguish ecosystem types by means of specific sets of species types or specific abiotic conditions (e.g. when distinguishing a lake ecosystem from a forest ecosystem as different ecosystem types). The classification here is in the eye of the beholder. It is one of different concepts, different *ideas* of what we consider to be the characteristics of an ecosystem. We could thus look at a specific area with its organisms, and investigate if it was a functioning ecosystem (or contained functioning ecosystems) from the perspective of these different ideas and the types derived from them. For the same area, the question of whether the ecosystems are functioning (properly) might then be judged completely differently, depending on which of the types was used as the background definition for an ecosystem. Some people would consider a corn field to be a properly functioning ecosystem as long as it delivered the services demanded from it (mostly those processes that provide food). Others would not call it a functioning ecosystem because it is dependent on human support (inputs of energy and substances), with no self-regulation.

So let's take a closer look at each of the types.

7.1.1 Ecosystem functioning and the generic-type ecosystem

The generic-type ecosystem is functioning as long as (any) organisms and their interactions with each other persist at all. Such a system thus corresponds to the generic definition of an ecosystem: all the organisms in a place together with their abiotic environment. The boundaries are normally set topographically and for convenience. There is hardly any way in which it can *not* function – unless all the organisms vanish. We thus have a kind of *generic* or *minimalist* concept of ecosystem functioning here. 'Functioning' in this case is synonymous with 'proper functioning'. Or, expressed differently: being devoid of stronger normative determinations,

Box 7.1 *The functioning of wilderness ecosystems*

In describing a wildness-type ecosystem, I have consciously been using the term 'wildness' and not 'wilderness'. While the latter refers mostly to a 'pristine' area that has not been used by humans, 'wildness' refers to the absence of deliberate human control, to 'free-running' ecological processes, no matter how 'pristine' the area is (see discussions in Cole, 2000; Chapman, 2006; Ridder, 2007). In this sense, every wilderness area is characterised by wildness, but not every wild area is wilderness. As wildness and wilderness play important roles in biological conservation strategies, a few more words on the notion of ecosystem functioning in this context are in order. The idea of wilderness is highly complex and contested (e.g. Nash, 1982; Cronon, 1995; Callicott and Nelson, 1998; Nelson and Callicott, 2008). Although 'wilderness' in principle could be understood as referring just to free-running ecological processes (uncontrolled by humans, i.e. wildness), the ecosystem idea actually associated with wilderness and 'letting nature take its course' in conservation practice is mostly much more specific. It is not just, as one might expect, *any* processes that are seen as constituting a (functioning) wilderness ecosystem, but frequently those that – without human influence – support a historical state that is deemed to be the *natural* one (e.g. Cole and Landres, 1996). The assumption is thus – as partly in the Yellowstone case (Section 4.3) – that ecosystems before humans influenced them were intact – they were properly functioning. As soon, however, as they deviate from this presumed pristine state – that is, from specific species compositions and/or distributions or physiognomies – their character as functioning wilderness ecosystems becomes doubtful for many advocates of the wilderness ideal. In Germany the wilderness idea was discussed under the label of *Prozessschutz* (process conservation). While the idea of letting ecosystem processes run free was generally supported, there was a clear distinction between desirable and undesirable processes and – not the least – the states resulting from them (Scherzinger, 1990; Potthast, 2000b). Again we see here how important the idea of naturalness is. Thus, the ecosystem within the practice of wilderness management (although this expression is in a sense a contradiction in terms) deviates often from the generic ecosystem definition and even from the wildness-type ecosystem. It can sometimes even come closer to the species-specific ecosystem type described below.

the parlance of 'proper functioning' is tautological. The qualifier 'proper' does not make sense here at all. This means, however, that there is also no specific reference state against which functioning could be assessed. It can just be *described* by various processes and the properties derived from them, without one of them being of more importance than others, without one state being privileged over others.

In a conservation context, a basic reason for the decision to make this a target system is to have ecosystems that develop without human control. In most cases this is applied to 'pristine' areas, but also, such as in densely settled areas of central Europe, to areas left to unaided successions. In this latter case a formerly managed forest (such as the Bavarian Forest National Park in Germany) or even a former strip mining area (see the case study in Section 7.3), if left unmanaged, can be a functioning generic-type ecosystem. In a conservation context, such an ecosystem would even be seen as functioning *properly*, but only if no human control is exerted upon it. Here, even the *exclusion* of some processes, namely active control by humans, is necessary to consider the ecosystem as functioning. But distinguishing between natural and anthropogenic processes is a *cultural* distinction and not pertinent to the generic definition of ecosystem functioning as such (see Box 7.1). With this cultural restriction of *permissible* processes we can therefore speak of a subtype of the generic type of ecosystem functioning as that of the 'wildness-type ecosystem'.

7.1.2 Ecosystem functioning and the process-focused-type ecosystem

All concepts of ecosystem functioning in some way refer to processes. 'Functioning' implies dynamics and interactions within the system. In this type, however, the processes are the very focus of the system definition. In a conservation and resource management context, it is often the processes (and their bearers) selected as ecosystem services that are of concern; they guide the idea of what constitutes a functioning ecosystem.

A process-focused type of ecosystem is considered to be functioning as long as it continues to provide the processes considered to be characteristic and desirable for the system. It is this kind of ecosystem type that dominates the understanding of 'overall ecosystem functioning' in the biodiversity–ecosystem functioning (BEF) debate (Chapter 3) and resilience research (Section 6.2).

Characteristic of this type is that the individual species are not of concern as such (in terms of their taxonomic identity), but only with respect to their *function* within the system, i.e. their *role* for the proper performance of the ecosystem processes in focus. These roles are called 'functional types' when the abstract grouping is meant, and 'functional groups' when individuals in a concrete ecosystem are dealt with (see the excursus in Section 3.2.3). In defining the ecosystem, functional groups and the processes they perform depend upon and reinforce each other. A functional group is only 'functional' with respect to a specific process and a process is connected to specific organisms that contribute to it or are affected by it. As a consequence, the proper functioning of the system may also be indicated by the presence of all required functional groups.

The system remains a functioning ecosystem even when some species are going extinct, as long as other species can take over their roles. This is the basis of the notion of 'species redundancy' (Walker, 1992; see also Box 4.2). Also, if this approach is taken seriously and in a strictly scientific manner, there is no difference between 'native' and 'exotic' species, as long as they perform the same roles.

A broad variety of more specific definitions exists for this type, differing in the choice of which processes – and so which 'functional types' of species – make up an ecosystem; I have already demonstrated this in connection with BEF research (Chapter 3). It is highly difficult to characterise the 'necessary' processes that make up an ecosystem. There is no unique solution to this problem, no clear and agreed-upon set of processes. While some scientists would, for example, say that the process of autochthonous primary production was a necessary requirement of an ecosystem (and thus the functional type 'primary producers' is a necessary part of an ecosystem), others would argue that cave ecosystems, with only allochthonous primary production, should be considered as functioning ecosystems in the sense of a process-focused ecosystem definition. The question is: what types of organisms and processes have to be present at minimum to speak of an ecosystem, and at what component resolution? Is it enough to have interacting groups of producers, consumers, and decomposers to make up a functioning ecosystem, are more fine-grained distinctions required within each group? Or must distinctions of a different kind be added, such as pollinators, seed dispersers, or ecosystem engineers?

Boundaries of such ecosystems may be defined either by topographic criteria, including purely pragmatic ones (e.g. administrative boundaries or boundaries of experimental plots) or by process relations (e.g. a watershed approach, which is mostly at the same time also a topographic boundary). Setting topographic boundaries is the most common approach here.

The loss of one of the functional groups, as a qualitative change, is also a loss of functioning; the ecosystem loses its completeness. Assessing the degree of ecosystem functioning on a quantitative level is only possible after reference conditions for the 'permissible' process rates have been defined, for example, the proper level of primary productivity (as a range, not as a fixed value).

Particularly, the expression of this ecosystem perspective in the form of the ecosystem-services type has many applications. Here, the functioning of the ecosystem is defined by the continued performance of those processes that support the provision of human benefits. There are many variations of the system (and its intactness) possible, depending on the specific services that are selected (see the excursus on the ecosystem services concept in Section 4.1.2). It is, by its very character, not restricted to unmanaged systems. 'Pristine' ecosystems are also viewed from the perspective of what benefits they provide to us (e.g. the tropical rainforest as providing climate-regulation, pharmaceutical drugs, or cultural benefits), but the systems we have traditionally defined or even constructed and engineered for providing ecosystem services are managed ones, such as agroecosystems. In fact, the distinction between natural and managed ecosystems is (or at least should be) not really relevant here. In consequence, naturalness is not an important background criterion for defining reference conditions of this kind of ecosystem.

As in the discussion about the wildness-type ecosystem, the important general question that comes up here is whether (proper) functioning is really given when the system retains its processes (and services) only with *continued human intervention*. Agroecosystems are the prototype here, but it is also relevant when restoring traditional culturally moulded ecosystems in Europe, which are important targets of classical conservation efforts, in Germany and elsewhere. Can we really speak of such systems as functioning ecosystems, even though they provide ecosystem services? If the characteristic of being self-sustained (which amounts to a rather high degree of required interactions on the I-axis of the SIC model) is necessary for defining a functioning ecosystem, agricultural systems could never be functioning ecosystems at all. These same would hold for many open-grassland systems used for recreation, species conservation, and/or fodder, which are maintained only by perpetuating or mimicking traditional land use methods.

In the ecosystem-services type, individual organisms and their aggregates are also primarily perceived with respect to their functions.

Box 7.2 *Ecosystem services and ecosystem functioning: some caveats*

Assessing the state of ecosystems via the services they deliver has become prominent through the Millennium Ecosystem Assessment (MA), which used ecosystem services as its major framework (see Section 4.1). Since then, ecosystem services have become a central concern of environmental policies and environmental research. There is sometimes the assumption that 'natural' ecosystems are able to deliver ecosystem services much better than those modified, impacted, or created by humans (Westra, 1998; Aronson *et al.*, 2007; Moss, 2007). While this may be true for some services, it is certainly not true for all. In particular, biomass production for food, one of the most valued 'providing services', will not be delivered in sufficient levels by unmanaged ecological systems to feed six billion people. There are trade-offs between the delivery of the various possible ecosystem services within an area. The provision of some services (such as biomass production for food) may preclude the provision of others (such as processes and products used for recreation) – with significant societal consequences. This, and also trade-offs with other potential conservation goals, should be kept in mind when promoting the idea of assessing ecosystem functioning via an ecosystem-services-type concept of the ecosystem. 'Ecosystem services' is a highly utilitarian concept. For this reason it has been heavily criticised as endangering conservation goals that are not use-oriented but that propagate (also) the protection of nature, or parts of it, for the sake of nature itself (McCauley, 2006; Ridder, 2008). At least if the notion of ecosystem services comes with the claim of covering all of what is important in nature conservation, this danger really exists. Some proponents of the concept would argue, however, that such concerns are unjustified because the ecosystem-services type also embraces processes leading to cultural benefits ('cultural services' in the MA terminology), in which all kinds of non-use values are integrated. In fact, the MA authors also emphasised that ecosystem services are not the only reason to protect biodiversity and ecological systems (Millennium Ecosystem Assessment, 2003, pp. 33 f.). To my knowledge, an in-depth investigation of the ethical implications of the ecosystem services concept and of its appropriate domain of application remains a desideratum.

The resulting functional groups/units have been called 'service providing units' (SPUs) and 'ecosystem services providers' (ESPs) (Luck *et al.*, 2003, 2009; Kremen, 2005) by some authors. The idea, as described by Luck *et al.* (2009, p. 225), is not focused on individuals or populations alone, but 'encompasses service providers across various organizational levels, from populations of single species to multifunctional groups and ecological communities'. The level of component resolution under which organisms are perceived is thus highly variable.

A good example of how a functioning ecosystem is considered to be constituted through the necessary functional groups for the provision of ecosystem services and the benefits associated with them is given in the study of Swift *et al.* (2004). In their paper, they suggest a nested (hierarchical) set of 'key functional groups' for an agroecosystem. They emphasise that their classification aims specifically at agriculture and related ecosystem services, and that no all-purpose classification of functional groups is possible.

Characterising ecosystem functioning by means of ecosystem services seems, on the one hand, easier than the more general approach of the process-focused type, as the desired (and/or required) ecosystem services are obviously things determined by humans. Identifying which processes are the decisive processes (and thus ecosystem services) that lead to the desired benefits for humans is, however, very difficult and ambitious. It is currently – and will be in the future – a matter of much research (e.g. Kremen, 2005; Kremen and Ostfeld, 2005; Luck *et al.*, 2009).

Sometimes, proper ecosystem functioning is seen as a *prerequisite* for the optimal delivery of ecosystem services. Obviously, proper ecosystem functioning in these cases is defined in some other way. We might, in general, also look at ecosystem services *after* we have defined an ecosystem and its functioning on other grounds (e.g. physiognomic ones or just the generic definition). We can then investigate which services and how much of each the system provides.

7.1.3 Ecosystem functioning and the physiognomic-type ecosystem

A physiognomic-type ecosystem is considered as functioning as long as its dominant life forms are perpetuated.

The (often implicit) definition of this type of ecosystem emphasises the dominant life forms of a system when it comes to describing system identity or assessing the maintenance or loss of ecosystem functioning. The boundaries of such systems are clearly set topographically. Processes

Fig. 7.2. Fredric Edwin Church's famous painting 'Heart of the Andes' (1859) followed Humboldt's ideas of a physiognomic description of landscapes. Having revisited the places of Humboldt's American journey, he wanted to depict the landscapes both scientifically accurately and with the aesthetic impression they conveyed by their physiognomic appearance. Copyright: bpk/Metropolitan Museum of Art, New York. Photo reproduced with permission.

are taken for granted and – in contrast to the functional-group type – are not the focus of defining the system; they are only the *mechanisms* for maintaining a functioning system defined otherwise. Instead, selected 'typical' species are decisive for defining the system and indicating its proper functioning. Even the major species may be substitutable as long as they belong to the same *life form*. The presence or absence of most other species does not matter, however, at all.

Physiognomic criteria are among the oldest criteria used to describe and delimit ecological units, starting with the concept of the plant formation as introduced by Alexander von Humboldt and August Grisebach (see Box 7.3). They predate ecological science as they also refer to everyday/colloquial distinctions between different parts of nature, as, for example, between forest, steppe, and fen.

Box 7.3 *The physiognomic tradition in ecology*

The physiognomic tradition within ecology dates back to one of the most influential antecedents of ecological science, namely Alexander von Humboldt. Reaching back to antiquity, physiognomic approaches were originally applied to humans, and were here used to infer the character of people; the physiognomy of a person was thought to mirror this character, its essence. Physiognomy has a quite chequered history, with frequent misuses, and received its last major boom during the eighteenth and nineteenth centuries, especially with the work of Johann Caspar Lavater (1741–1801), who tried to develop it into a serious science. Humboldt, who first systematically applied physiognomy to plants, groups of plants, and landscapes, was influenced by Lavater. Humboldt described the physiognomy of landscapes (its total impression; German: *Totaleindruck*) by means of morphological plant forms (Humboldt, 1807 [1969]). His idea was, however, not strictly ecological in the modern sense. The physiognomy of the landscape, for him, always had a highly aesthetic element, explicitly related to art (Hard, 1969; Hoppe, 1990). It pointed, as Lavater's physiognomy of humans, to a higher reality of nature, which could not be grasped by reason alone: the inner structure of the object, its unity in diversity. Humboldt at first distinguished 16 different plant forms on the basis of morphological, not taxonomical groupings. The physiognomy of the landscape, for Humboldt, derived from the arrangement and gross abundances of the dominating plant forms. Humboldt's ideas formed the basis for subsequent plant geography and early ecology. Building on a further development of Humboldt's physiognomic plant forms, in 1838 August Grisebach described *plant formations* – a term coined by him that became highly influential in the beginnings of ecology for classifying and mapping vegetation. While Grisebach still retained some of Humboldt's aesthetic and emotional elements, later authors (e.g. Oscar Drude and Eugenius Warming) 'purified' the physiognomic approach almost completely of these aspects and transformed it into a purely ecological concept, not the least by emphasising correlations between the dominant life forms and ecological factors. The formation was thus seen as 'an expression of the prevailing living conditions (climate, soil, mutual relations of organisms)' (Flahault and Schröter, 1910, p. 5; my translation). It turned out, however, that such correlations were difficult to establish unequivocally. In consequence, the formation as a physiognomic ecological unit lost importance, in favour of those units that were determined either by taxonomical or physiological criteria, such as the plant association. As an explicit criterion, physiognomy became rather inconspicuous within ecology. Only as a rather coarse scale description and classification of vegetational units is the formation concept still in use.

This physiognomic-type ecosystem is often in people's minds when empirically distinguishing ecosystems in an area, and when ecosystems are classified and mapped. It is likewise frequently used as a guiding perception when ecosystems are said to become dysfunctional, degraded, destroyed, collapsed, or turned into another ecosystem. The now established use of 'repeat photography' for comparing historical and present states of ecosystems (e.g. Meagher and Houston, 1998; Higgs, 2003) emphasises, by its very macroscopic nature, this view of ecosystem functioning. That is, a 'degradation' of the ecosystem is being inferred through a change in its physiognomic appearance. Although 'physiognomical' criteria have often been considered as old-fashioned, too coarse, or even unscientific, they implicitly play a very important role within ecology. There are many examples of this. One is the definition of a savannah as a (tropical or subtropical) mixed tree/shrub–grass community. Jeltsch and colleagues, for example, write: '[E]daphic, topographic and many other conditions vary widely among savannas, and in particular the set of species and processes involved in the dynamics of any savanna is unique. But despite all these peculiarities, we usually have no problem in identifying a given tree–grass mix as savanna vegetation' (Jeltsch et al., 2000, p. 161).

The change of the savannah into either a woodland or a grassland is then considered as a change of ecosystem type. Although there is still an ecosystem in the sense of the generic-type ecosystem, it nevertheless ceases to function as a savannah ecosystem. Likewise, as noted above, we mostly do not consider a forest as dysfunctional or collapsing when 'only' understorey plants, some insect species, or even birds and large mammals (e.g. wolves and bears in western Europe) go extinct. But we perceive it to have degraded or collapsed if the trees are affected and die. Even the replacement of one tree species through another (e.g. exotic species) might not be perceived as a loss of functioning of a forest ecosystem. Other examples exist for aquatic systems. The shift between alternate states of lakes is (also) a physiognomic one: the dominant life forms changing between a clear macrophyte-dominated state and a turbid microphyte-dominated (i.e. plankton) state (see the case study in Section 5.1). The shift between different states of coral reefs (coral-dominated to algal-dominated state) is also a change in physiognomy. The physiognomic type of system is the one which is mostly referred to when thresholds are described for ecosystem changes. It is not changes in flow rates (of energy, nitrogen, etc.) as such that are used to decide when an ecosystem has significantly changed. The threshold is seen as crossed only when (also) the physiognomy of the system has changed. However, a shift in physiognomy normally also *indicates* major shifts in many ecological processes within the boundaries of the system.

The approach of assessing ecosystem functioning via the maintenance of the physiognomic appearance of a system has both advantages and problems. On the negative side, the physiognomic approach leads to a rather coarse definition of ecosystems. There is no unambiguous correlation between physiognomy and the processes that create it. Physiognomic boundaries and physiognomic appearances of ecological systems can be caused by a variety of different factors, ranging from discontinuities in abiotic conditions, over historically contingent events (such as natural or anthropogenic disturbances), through biotic interactions which reinforce physiognomically defined ecosystem boundaries. Also, there are many different ways of classifying physiognomies. Where do we set the limits between classes? Moreover, where would we say that a physiognomic boundary constitutes the boundary of an ecosystem or where is it just the boundary of a patch within a larger system? These latter questions are, however, not unique to a physiognomic-type definition of ecosystem functioning.

On the positive side, physiognomic-type ecosystems are, in spite of the problems just described, comparatively easily delimitable in space. Also, shifts in physiognomy are very conspicuous, and, as said above, mostly indicate concurrent major changes in ecological processes. Finally, they provide good connections to the aesthetic dimension of nature conservation.

Nevertheless, perceived as desirable or not, we must acknowledge as a matter of fact that the physiognomic-type ecosystem perspective on ecosystem functioning is present in ecology and conservation.

7.1.4 Ecosystem functioning and the species-specific-type ecosystem

A species-specific-type ecosystem is considered as functioning as long as all species living in it at a specified reference date (or which are considered as typical for it) persist over time.

Boundaries are defined topographically, but may sometimes also be defined via the home ranges of the largest species (and thus process-based).

Such a type is often the goal of national park management and is what traditional conservation, as species protection, would frequently see as the evidence of a functioning ecosystem. 'Saving all the parts' (here: species) may either be seen as a goal in itself (caring about all species as such), as a means to conserve a natural or historical state, or simply as an *expression* of a functioning ecosystem (where the 'ecosystem' is in focus).

Note, however, that in some cases all ecological *processes* are also to be maintained, as in the management of Yellowstone National Park (Section 4.3). That is, the number of selected elements on the selected phenomena axis of Fig. 7.1, which includes both objects *and* processes, is even higher than in many traditional conservation approaches, where processes are just a *means* to realise ecosystem functioning, but not a necessary criterion for assessing its proper functioning.

Many approaches of ecosystem integrity (Section 6.2) come close to this kind of species-specific definition for ecosystem functioning in their practice – using a specific species composition as an indicator for a functioning ecosystem.

In the strict sense, the functioning of this kind of ecosystem would be impaired as soon as *any* species included in the specification of the reference state was lost. In practice, all species within the boundaries of an ecosystem are not really known, nor would *any* loss of species (or addition – as by exotic species) be considered as being problematic. There are always 'transient' species, and changes in time and space. In practice, the reference state either refers to those species deemed typical (or natural) for the area and/or habitat type, or it is 'downgraded' to a physiognomic-type ecosystem. It is evident that *selections* on the basis of previous knowledge, educated guesses, or more specific conservation aims have to be made, which in themselves have to be justified by extra-scientific arguments.

7.1.5 What is the use in describing different ecosystem functioning types?

One might ask if these different ecosystem types are not just different aspects of the same ecosystem, like in the famous parable of the elephant and the blind men, where each man touches a different part of the animal (such as the ears, the tail, or the tusks) and comes to a different interpretation of the object, the elephant. The answer is: yes and no.

Yes, because we can apply these different perspectives to the same chunk of nature – for example, an area of land in the Bavarian Alps. We could describe this area with respect to the processes that go on there, with respect to the services and benefits that can be derived from the land, with respect to the dominant life forms, and with respect to the species present. But – and here comes the 'no'– if we perceive of this area as an ecosystem (or several ecosystems, or parts of ecosystems), the boundaries of the different systems created by different definitions (ecosystem types) would not match, and even less could we reach a clear and unambiguous statement of whether or how properly the ecosystem(s) is/are functioning. This is clearly dependent on our choices with respect to the precise system definition and with respect to the reference state. These choices, in turn, are based on our interests, be they scientific, individual, or societal. The *totality* of nature in a place cannot be assessed – not because we have only insufficient data or data processing capacities, but in principle. It is not possible. As a consequence, we will never be able to develop one unified ecosystem theory and thus a unified theory of ecosystem functioning. We might, however, be able to develop good theories for more restricted domains of the ecological universe. Such theories might apply for some more specific types of ecosystem definitions, but nevertheless not treat each definition as a unique case. We thus have to find a middle ground between idiosyncratic ecosystem definitions and an all-in-one ecosystem theory. The types might be similar to those described above, they might also be different. This is just a first proposal. It is also necessary to search for empirical correlations between the performance of different types. They are clearly not completely independent from each other. But even if we do not have clear-cut ecosystem theories for different types of ecosystems and their functioning, we can nevertheless *communicate* clearly about them and find ways to assess in a reasonable way if and to what degree they are functioning. This is, as has been emphasised already, an enterprise at the interface of science and society. The next section provides some guidelines on how to proceed in assessing ecosystem functioning.

7.2 Some guidelines for conceptualising and assessing ecosystem functioning

Questions we face when it comes to putting ecosystem functioning concepts into practice are: how do we get a grip on ecosystem functioning, given that there seem to be so many possibilities for conceptualising and measuring it, with no preferred model in sight? Who decides and how? How can we avoid statements about ecosystem functioning that are completely arbitrary, not comparable and/or useless in the practice of ecology, conservation, and resource management?

The concept of ecosystem functioning poses problems to science: the plurality of existing concepts, the fact that ecosystem functioning is not a purely scientific concept, and also that even ecosystems are not simply given by nature as 'natural kinds'. But if we acknowledge these findings, the problem can be turned into an advantage for ecology, conservation biology, and other application fields. First, it relaxes any possible search for the 'true' definitions of ecosystems and ecosystem functioning in favour of a search for (several) *appropriate* definitions. Second, and even more importantly, it opens up a chance to be an interface between scientific descriptions and societal choices, not mixing them, but linking them.

A couple of decisions or *choices*, both scientific and individual and societal, have to be made to conceptualise and assess ecosystem functioning. I will explicate these choices below. Some of them have been discussed already in previous chapters, some of them are added only here.

We will have to think about the *methodologies* in order to make these decisions. This pertains especially to those choices that are not only a matter of scientists and their community, i.e. when it comes to linking science and (the other parts of) society. This is necessary in many cases, as I have shown above, and is anything but trivial. Here we reach the field of social science methods where social scientists and/or interdisciplinary groups including ecologists and social scientists must come together to apply and, where necessary, even develop appropriate methods.

What I present in the following is a procedural approach, expressed as a kind of 'checklist' of things to be considered for the assessment of ecosystem functioning. First, I will describe the major choices that have to be made, and then I will present some procedures that are relevant for arriving at these choices, as a scientific *and* societal enterprise.

7.2.1 Choices

Assessing whether or to what degree an ecosystem is functioning requires choices. This applies also to determining what ecosystem functioning means, and *how* ecosystem functioning should be assessed. All data that we may assess become only relevant and gain only precise meaning in the context of the conceptual framework specified by these choices.

From a scientific perspective, four major choices have to be made, namely about:

(1) an appropriate ecosystem definition;
(2) a reference state or reference dynamics of the ecosystem;

(3) the specific measurement variables; and

(4) the spatial and temporal scales under which the system is to be observed.

These four scientific choices, however, are determined by individual and social choices. If they are to be scientific and/or socially relevant choices, they should be rational ones, the reasons for which should be made explicit and thus open to discourse. A fifth and foremost choice, preceding the more scientific ones and being the one which determines everything else, is about the *purpose* of the respective study dealing with ecosystem functioning.

Purpose of the study: who makes the choices and why?
The first decision that must be made is about the purpose of the study, and thus the purpose of conceptualising 'ecosystem functioning'. This is, by necessity, a normative decision, a value decision. It refers to *interests*: those of the individual researcher, the scientific community, the funding agencies, of society at large – represented by individuals and specific groups who have an interest in the study (stakeholders). Such stakeholders may be, for example, politicians, administrations, environmental agencies, interest groups from agriculture or tourism, environmental NGOs, but also different groups of local people who are affected by policies or management plans based on a notion of ecosystem functioning. The degree to which stakeholders beyond scientists (although scientists also have stakes!) must be involved varies, mirroring different degrees of normativity – from mostly methodological to morally relevant values (see Section 5.2). If the study is only of academic interest, without much relation to applied issues, it is up to the scientist and his or her scientific community to decide about the other items. There is, then, a broad range of possible conceptualisations of ecosystem functioning, which nevertheless has to be made clear and explicit in order to allow useful comparisons, testing, or further generalisation (i.e. transfer to other systems). The other four choices to be made for conceptualising ecosystem functioning depend on the specific scientific questions, their broader theoretical context, and the assumptions made within this context.

If the results of a study on ecosystem functioning are, however, of an applied character, the specific societal purposes narrow down the appropriate definitions quickly. The ideas behind, for example, different conservation aims have to be analysed and 'translated' into definitions of ecosystems and their reference conditions. This does not always, but frequently does, involve active cooperation with stakeholders, especially in determining reference conditions. The idea of what a (functioning) ecosystem is, is often only vaguely recognisable and needs careful theoretical analysis and/or conversations with stakeholders, be they politicians, national park managers, or people involved in some local ecosystem restoration project. The vagueness of concepts should be taken as an opportunity to state more precisely the researchers' own notions of a functioning ecosystem and those of other interest groups. If the different concepts of different individuals and groups do not match, this should be the starting point for negotiations about aims and about the appropriateness of the different concepts, both with respect to the purposes they should serve, and with respect to their relevance to physical reality.

There are various reasons why one concept of ecosystem functioning is preferred over others: they may be guided by personal curiosity or (environmental) concern, scientific fashions (what counts as a hot topic, what is likely to get funded and published), what is considered as morally relevant (species, ecosystem services, aesthetics), either for its own sake or for the interests of humans (or at least specific groups of humans), etc. This has been shown in several of the case studies within this book (see also the description of ecosystem functioning types in Section 7.1 and discussion of the moral relevance of ecosystem functioning in Section 7.4).

Definition of the ecosystem
Defining the ecosystem concept is necessary to proceed from the intuitive notion of what an ecosystem is (which almost everybody has) to an expression that is specific enough to allow judgements about empirical states of the system. It should allow for statements such as that it has ceased to exist as an ecosystem or as *that* ecosystem (i.e. has turned into another type or alternative state). This expression should be precise enough to be intersubjective, that is, it should allow other persons who use the same definition to delimit the same physical object as an ecosystem in some place. Even the generic definition may be precise enough – if that is what is really meant! As described in Section 7.1, under this definition any system consisting of organisms and their environment would also be a (properly) functioning ecosystem.

As one methodology for defining the ecosystem and its reference state, the SIC model (see Section 4.3) is a useful and illustrative tool, and I will thus refer to it time and again in the following text. Other ways of unambiguously defining ecosystems might of course serve the same purpose.

Even though extra-scientific criteria determine much of the *selection* of an ecosystem definition, the definition of the ecosystem itself should refer only to scientific, ecological variables. Decisions as to what is deemed 'natural' or 'typical' influence the selection of the ecosystem definition, but they must not be part of it. For example, decisions about the appropriate ecosystem definition might also include a statement as to whether the ecosystem must be self-sustaining. This is relevant for deciding whether only an ecosystem working without human interference is perceived as being (properly) functioning, or if a system is also considered as functioning when humans (have to) manage it by providing it with additional resources or selecting species (e.g. agroecosystems). In the SIC model, the property of being self-sustained (or self-regulated) would be understood as part of the axis of internal relationships (I-axis), without being a cultural property as such. All the axes of the SIC scheme only refer to scientific, ecological variables. The distinction between the ecological and the social dimensions of ecosystem functioning is thus maintained.

In principle, a clear and unambiguous ecosystem definition could be given specifically for each single investigation, rendering each ecosystem definition unique. This would, however, not be useful in terms of scientific progress. It is necessary to arrive at a restricted subset of different definitions with an intermediate specificity, which might then allow some generalisation for specific *classes* of ecosystem definitions (expressed, for example, as segments of the SIC cube). The typology of ecosystem functioning concepts given above (Section 7.1) is a first effort towards such a classification.

The definition of the relevant ecosystem concept alone is not sufficient for an assessment of ecosystem functioning. To really measure functioning (in yes/no terms or as a gradual measure of proper functioning) it is also necessary to define reference conditions for the system (see also Box 2.2).

Reference conditions
Here, I use the phrase 'reference condition' to denote and include both reference *states* (variables remaining constant) and reference *trajectories* (defined changes of variables, e.g. successional or cyclic ones).

The reference conditions provide more information about the desired specific values of the variables seen as characteristic of a functioning system. Reference conditions have to relate to the selected variables of the ecosystem derived from its former definition. Reference states may be drawn from empirical data or from models. They are frequently taken from historical data or from contemporary systems that are seen as reference systems (i.e. considered to be functioning properly) for the ecosystem in focus (see the case study on the European Water Framework Directive (WFD) in Section 6.2). Defining reference conditions also involves clarifying the 'permissible' variability of the relevant variables. If the ecosystem moves beyond the range of these values, it either does not exist any more or – depending on the specific definition – exists in a degraded or damaged state, or functions less or not properly. Thus, 'forest degradation' (as a loss of proper functioning of a forest ecosystem, but not as its complete destruction) may be defined by specific intervals of values for crown closure, carbon stocks, or other variables (e.g. Sasaki and Putz, 2009).

The variability 'allowed for' in the reference conditions is also necessary because ecological systems of all kind are dynamic. Many things in ecosystems change all the time. Organisms get born, move, use resources, grow, interact, reproduce, die. Within the definition of an ecosystem, this is accounted for in different ways, depending on the definition: some of the highly dynamic variables may not be considered as relevant for assessing ecosystem functioning; they thus do not enter the definition at all. In other cases, such variables are part of the definition, but are given a large allowed variability of the reference condition. In yet other cases, the reference conditions for such variables may be described by specific trajectories (and also cyclic ones).

If thresholds for changes between the states of ecosystems (alternative stable states, domains of attraction, loss of functioning; see the case study in Section 5.1) can be determined, they must also be named here. If such thresholds exist, they greatly increase the usefulness of the model.

Measurement variables
The definition of an ecosystem, its reference condition, and the scale do not in all cases automatically prescribe by which variables ecosystem functioning, as specified by them, are measured. As has been discussed above (see especially Chapter 3), proxies or indicators for ecosystem functioning are frequently involved, such as a few *selected* variables like productivity, biomass, or the retention of selected nutrients as indicators for an ecosystem defined mainly in terms of biogeochemical processes. Also, processes are often indicated by their outcomes (e.g. productivity is estimated by using biomass as a proxy). The question that has to be

Box 7.4 *Resilience as a measure of ecosystem functioning*

As a measure for ecosystem functioning, normally constancy of some variables or a defined trajectory are applied, as characterising the persistence of the system over time. This is valid for most of the ecosystem definitions above. We may, however, also add resilience to disturbances as an additional measure for ecosystem functioning. Perceiving resilience as a necessary criterion for proper ecosystem functioning in general is much more ambitious than dealing just with system persistence. Many systems that may be described as persistent over time may not be resilient at all. If the measure of proper functioning is ecosystem resilience, a clear definition of the possible *disturbance(s)* to which the system should be resilient is also necessary (see Section 6.2). Depending on the background theory embraced, one may even extend functioning to the condition of the system following an 'adaptive cycle'. It is – again – a matter of perspective whether we include all stages of an adaptive cycle together as a sign for the ecosystem to be (properly) functioning or view each stage separately. Assessing the proper functioning of the individual stages then requires setting up individual reference criteria for each of them. The idea of ecosystem functioning – in the sense of degrees of proper functioning – becomes highly complicated here. Please note that it is not possible to display resilience (which is a *potential* of the system, and not just an actual condition) directly in the SIC model.

scrutinised very carefully, however, is whether the variables measured are in fact sound proxies for those variables they are about to represent.

Proper functioning of an ecosystem in general is indicated by the *constancy* of the selected measurement variables over time or *defined trajectories* of these variables (within the respective defined intervals of reference conditions). More seldom, resilience to disturbances is used as a measure for functioning (see Section 6.2 and Box 7.4).

Scale

An ecosystem may be perceived on many different *spatial scales*, from a few grams of soil or a water-filled tree hole with its organisms, up to the whole biosphere. While it may be considered as not functioning on a small scale, it may at the same time be perceived as functioning when viewed on a larger scale, i.e. perceived as part of a larger ecosystem. A boreal forest ecosystem (physiognomic definition) might be considered as not functioning any more when it becomes burnt through a wildfire. On a larger scale, where the burnt forest is seen as only part of a much larger forest ecosystem, even large-scale disturbance events such as these might be considered under a patch-dynamics perspective, with wildfire then being perceived as part of the forest dynamics that even *maintains* the overall functioning of this ecosystem (e.g. Attiwill, 1994). The question is: at what scale do we perceive of an ecosystem and its functioning? At the scale of a stream reach, a whole stream, a whole watershed, or even larger units (e.g. a whole island like Navarino Island (Chapter 2) or a large area like the Greater Yellowstone Ecosystem (Section 4.3))?

While the ecosystem as such is a scale-insensitive *concept* (it can be described on scales as small and transient as a plankton patch up to the scale of our living planet as a whole), scale becomes relevant when it comes to assessing the functioning of an ecosystem in specific empirical settings. Thus, a decision about the scale on which the ecosystem is to be investigated is crucial for any clear statement about its functioning.

Assessing ecosystem functioning is also sensitive to the *temporal scale* under which a system is perceived. It requires a specific minimum time in which the defining characteristics of the system do not deviate more than allowed from the reference state or reference trajectory. If the system is functioning properly only in the very moment we investigate it, the statement may often be trivial and/or useless. If, at the other extreme, we extend the temporal scale too much, almost no system would be judged as functioning, as there are always larger changes during geological timescales, due to evolution, climate changes, etc. The appropriate scale depends on the system (e.g. the turnover rate and size of organisms), but also on the purpose of the investigations. With respect to stability measures (here: persistence), Connell and Sousa (1983) suggested that the temporal scale of an investigation should at least extend beyond the generation time of the longest-lived species, or in their words, 'one turnover of all individuals of that species in the place' (p. 791).

Ecosystem functioning might thus be thought of as a snapshot of the condition of an ecosystem or as a goal to be retained for a longer period (e.g. in the context of the sustainable use of resources). To some degree it always reaches beyond the momentary condition, but the question is, how far should it or must it reach? Systems may also be exploited for a while and then cease to function on behalf of this use (such as many agricultural systems in the tropics) – something which is almost never desirable. They may persist as functioning with or without external support (gardens, agricultural ecosystems).

7.2.2 Procedures

Participation

Participation has become an important issue in conservation biology and environmental management. There is a large literature about this issue (see Webler *et al.*, 2001; Stoll-Kleemann and Welp, 2006; Dietz and Stern, 2008; Reed, 2008 for overviews, but see Cooke and Kothari, 2001 for a critique of many approaches). It would go far beyond the possibilities of this book (and my own competencies) to address it here in any detail. A few words, however, are nevertheless necessary, at least to sketch the major questions involved.

As noted above, participation is not always necessary when determining an appropriate way of conceptualising and assessing ecosystem functioning, especially in a purely scientific research context. For many applied purposes, in conservation and natural resource management, it is, however, desirable or even necessary.

I am using 'participation' here in a broad sense. That means it does not only include the participation of different groups of the local population, but also of decision makers (administration, politicians), i.e. the policy side. It therefore also involves deliberative processes. Important and non-trivial questions are: which interests are at stake? How are the stakeholders selected (e.g. Ravnborg and Westermann, 2002)? Who selects them, on what basis, and with what legitimacy? What is the aim of the participation process? By what methods can stakeholders be involved? How can differences in education and power be accounted for? How can stakeholders be motivated to take part in participative processes?

While participation is common in many forms at the stage of general goal formulation and implementation of scientific results in ecological conservation, management, or restoration, the plea made in this book is to use participative approaches *in the phase of concept clarification and concept building*, i.e. when determining what is meant by ecosystem functioning and how it is to be assessed. The reason is that our concepts of ecosystem functioning and other ecological concepts (e.g. biodiversity) *have an influence* on management and conservation goals. These goals, once they are implemented, influence people's lives (and the lives of other organisms) and are thus morally relevant. But, as has been described in this book, they are also not independent of people's values and norms.

Analysing and negotiating the variety of stakeholders' concepts has been done frequently for different concepts of 'nature' (e.g. de Groot and van den Born, 2003; Berghöfer *et al.*, 2010), but less for more scientific concepts (but see Fischer and Young, 2007 for 'biodiversity'). Ideas of nature and of what is 'natural', as has become clear at several points within this book, have a strong influence on concept formation in ecology and conservation biology and on the practice of these disciplines. As concepts of nature are mostly also an implicit part of people's everyday experiences, they might be good starting points for approaching ecological concepts such as biodiversity and ecosystem functioning. However, for this purpose, *translation work* is necessary.

Translation

Translation is part of the participation processes. It is also required if the aim is only transmitting (uni-directional) information about ecological issues; but this is not what I discuss here. Translation is thus embedded in a specific participation context. It is only this context – for example, to develop better ecosystem management strategies – that will, in most cases, motivate people to take part in a participation process. This holds true even more when the task is one that, at first sight, is as abstract as translating between scientific ecological concepts and people's everyday perceptions of the world. Participant motivation is a problem in many participation and valuation processes, which is easily forgotten by researchers, mostly engaged as they are in their own research. We should deal in a very careful and honest way with the expected and possible benefits that participation processes can have for those involved.

The question of how far we can translate between ecological concepts and lay concepts, such as the concepts of nature discussed, is a difficult one. It is also not evident in which instances such translation is really necessary (beyond communication of scientific results to 'the public'). Certainly, there can be no 'one-to-one' translation. We know that even translation between different natural languages (such as English or German) is possible only to a limited degree. Translation is more difficult the further removed from scientific thinking the people to be involved are. Also, it is not enough to just position lay knowledge as either 'right' or 'wrong' with respect to 'objective' scientific knowledge.

Specifically for hybrid concepts such as biodiversity, resilience, and ecosystem functioning, which are highly dependent on normative choices, efforts to translate between both realms can be very important. This is still a challenge. Approaches like the SIC model (from the scientific side) can aid in this process. The simple analyses performed with the model in Section 4.3 (intactness of the Yellowstone ecosystem) and Section 7.1 (typology of ecosystem functioning concepts) show how this may be approached, here mostly on the basis of using published sources, i.e. without active stakeholder involvement. Other approaches also use narrative (local/traditional) knowledge (e.g. Berkes *et al.*, 1998) or interactive approaches like interviews and focus groups (Fischer and Young, 2007; Berghöfer *et al.*, 2010). There are no generally established methods for this purpose. Much methodological development is still needed to translate between the scientific ecological sphere and that of everyday concepts, experiences, and knowledge.

The final step: assessing ecosystem functioning
Having defined the above variables (ecosystem definition, reference condition, measurement variables, scale), the final step appears to be quite simple, at least with respect to theory (it may be very laborious and ambitious in terms of the empirical methodology). Assessing whether an ecosystem is functioning (properly) in principle simply means to compare the predefined condition (state, range of states, trajectory) to the actual condition of the system. It mostly not only refers to the very moment, but to its (projected) future, or more generally a comparison between two instances in time (see Jax *et al.*, 1998 for an application of the SIC model in this context).

To predict the future development of ecosystems and their functioning is a highly ambitious task, to which much ecological research is devoted. Finding more generally applicable answers, however, first requires a clear understanding of different ideas of ecosystem functioning.

Of the research and application fields for which ecosystem functioning plays an important role, restoration ecology is coming closest to an ecological (or at least ecology-related) discipline that works at the interface between science and society. As such, the last case study of this book is devoted to this research field, because it allows a rather comprehensive exemplification of the issues discussed in the previous chapter.

7.3 Case study: ecological restoration and ecosystem functioning

I still remember an evening in the early 1990s, when Gerhard Wiegleb, a colleague of mine from the University of Cottbus, took me on a trip to one of the vast post-mining areas in Lower Lusatia. He and his co-workers were studying succession on some of the dumps from the former lignite mining. Nearby, there were areas with planted trees, open succession areas with sparse spontaneous grassland vegetation, and newly forming lakes which slowly filled with the groundwater that was allowed to rise again after mining activities in this area had ended. On the banks of the lake, below some steep and unstable sandy cliffs, a flock of large birds (cranes, if I remember correctly) could be observed, and as the sky became red in the dwindling evening light, I could not help but feel more that we were in a wild and almost sublime area in the American Southwest than one of the most devastated landscapes of central Europe. In a way, my feeling of experiencing a wild landscape was not so wrong, at least if we do not connect wildness with the attribute of being in a 'pristine' state, i.e. with wilderness (see Section 7.1.1, above). At least in parts of the former mining area, nature was left on its own, without the use of stronger activities to mould new or recreate old landscapes. All in all, however, post-mining areas, at least in Europe, are a major object of active ('interventionist') restoration efforts, and in some countries, such as Germany, mining law even obliges mining companies to restore (or reclaim) these areas after mining activities have ceased. While this was done in the beginning as a kind of gardening or landscape architecture activity, it has, with the rise of ecology and especially of restoration ecology, increasingly become an enterprise that is performed on the basis of scientific research and advice. As such, it became a field for experimenting with ecosystems and their (re)construction, prompting Anthony Bradshaw, one of the pioneers in mine-site restoration, to speak of restoration ecology as an 'acid test' for our understanding of ecosystems, 'because each time we undertake restoration we are seeing whether, in the light of our knowledge, we can recreate ecosystems that function, and function properly' (Bradshaw, 1987, p. 26).

Restoring ecological systems is a rather old activity. In Germany, for example, the first forest was systematically replanted (seeded) in the fourteenth century (near the city of Nürnberg). In the Eifel highlands, the area where my family has lived for many generations, the new Prussian government, during the early nineteenth century, replanted large areas of the then devastated forests, although mostly with Norway Spruce (*Picea abies*), and not with the 'naturally' occurring deciduous tree species of the area (my grandfather still spoke of spruce as the '*Preusseboom*', the 'Prussian Tree'). Using conifers instead of the native deciduous trees was, however, not in the first place due to the effort of maximising timber production, but because many soils were thought to be so degraded that only spruce could be planted successfully, planned as an intermediate state for restoring the deciduous forest again (Schwind, 1984, pp. 144 f.). In the early twentieth century, the 'Dauerwald' (i.e. perpetual forest) concept of Alfred Möller even paved the way to a more 'ecological' way of restoring forests (see Schabel and Palmer, 1999). While the *practice* of ecological restoration is thus rather old, *restoration ecology*, as a scientific discipline, is still young, as is restoration that is guided explicitly by ecological principles. The major society of this field, the Society for Restoration Ecology (SER) International, was only founded in 1988.

Ecological restoration today covers a large variety of activities, ranging from the restoration of almost-pristine areas, through to restoring 'degraded' ecosystems, to the recreation of large-scale ecosystems in places where they have been completely destroyed, and sometimes literally turned upside down, as in many mining areas all over the world.

The purposes of restoration are likewise quite diverse, and have changed and broadened over time (e.g. Hobbs and Norton, 1996; Ehrenfeld, 2000; Higgs, 2003). They range from the purely technical task of stabilising landscapes by re-vegetating devastated areas, to rehabilitating the services provided by the former ecosystems, protecting biodiversity or even single species (including their reintroduction), to restoration as serving even a spiritual purpose, namely the reconciliation with nature – the process and the building of harmony of nature being an aim in itself. Not all of it (especially not the first of the purposes mentioned) would most likely be termed *ecological* restoration by many restoration ecologists.

The definition of ecological restoration as given by the SER International today is: 'Ecological restoration is the process of assisting the recovery of an ecosystem that has been degraded, damaged, or destroyed' (Society for Ecological Restoration, 2004; section 2).

Restoration here (and also in other definitions) is thus expressed in terms of restoring '*ecosystems*', not just single features of communities, ecosystems, or landscapes, even though the latter aims are generally not precluded (Ehrenfeld, 2000).

The goals of restoration are of course closely related to the idea of what good restoration is and to how the success of restoration measures can be assessed. To give an indication of what a successfully restored ecosystem looks like, in their *Primer on Ecological Restoration* (2004) the SER provide a (much-discussed) list of nine 'attributes of restored ecosystems'. It includes, among others, the restoration of a characteristic species composition and community structure, resilience to 'normal periodic stress events', the ability of the system to be self-sustaining, and not the least the normal functioning of the system. This is quite an ambitious list, although most authors acknowledge that not all of these attributes must be fulfilled to speak of a successfully restored ecosystem.

Box 7.5 *'Restoration': some terminology*

Following the generic dictionary definition of 'restoration' (e.g. from the *Oxford English Dictionary*), I am using the term 'restoration' here as an overarching concept, namely all those intentional actions aiming at bringing an ecological unit back to a previous, original, or 'normal' condition. There are a couple of other terms used in related meanings or synonymous to 'restoration', such as 'recovery', 'rehabilitation', 'reclamation', 'regeneration', 'ecosystem repair', 'remediation', or 'revitalisation'. They differ especially with respect to the starting point of the management (e.g. completely destroyed or only degraded), and their goals (e.g. restoration of structure or process, historical conditions, or ecosystem services), but no generally agreed-upon distinctions exist. For example, the difference between ecosystem restoration and rehabilitation is often seen in that the first refers to achieving both the structure and the processes or services ('functions') of the original systems, while rehabilitation only recreates the processes and/or services of the reference system (King and Hobbs, 2006). See also Whisenant (1999), Bradshaw (2002), and Zerbe and Wiegleb (2009) for the terminology of the field, and especially Higgs (2003), who provides a very good account of the conceptual difficulties of delimiting (ecological) restoration from other practices.

To specify these goals, and later, to assess if restoration has been able to reach or approach them, reference systems are needed (as demanded by Aronson *et al.*, 1993, 1995), or statements about the minimal properties that a system should fulfil (e.g. Jordan *et al.*, 1987; Hobbs and Norton, 1996). The restoration of ecological units is sometimes likened to the restoration of an old painting that is degraded but has basic features that are still visible (Aronson *et al.*, 1993). The same goal is often sought for ecosystems: the basic features of these units are targeted for maintenance or restoration. There is considerable disagreement, however, about what these basic features are. While some scientists try to restore ecosystems on a highly abstract level based on concepts such as flows of matter and energy (e.g. Brown and Lugo, 1994), others consider that it is not sufficient just to restore these processes, but that restoration of particular species or even local genotypes is necessary in order to capture the 'essence' of the ecosystem (Ashby, 1987).

In contrast to many areas of conservation biology, where the aim is protection of ecosystems in their *present* (mostly 'natural') condition, the case in restoration ecology is different, requiring more elaborate decisions about goals to be taken. In some cases (as in mining areas or new landfill areas), there is hardly anything left from the original ecosystems, and even the abiotic settings (e.g. soils or substrates) have been altered dramatically. So there are decisions necessary as to what kind of future ecosystem an area should be restored to, with several options existing in most cases.

The aim of restoring ('fully') functioning and self-sustaining ecosystems has been emphasised by many authors (e.g. Bradshaw, 1987; Harris *et al.*, 2006; Aronson *et al.*, 2007). Within the SER 'checklist', criteria #5 in particular explicitly refers to ecosystem functioning: 'The restored ecosystem apparently functions normally for its ecological stage of development, and signs of dysfunction are absent.'

Ecosystem functioning is obviously not the only goal of restoration, but an increasingly important one. It can, as implied in the SER *Primer*, be understood as one of many criteria for a successfully restored ecosystem, or it might be considered as the overall target description of restoration for which the other criteria are only specifications and/or which overrides other possible goals.

So, if ecosystem functioning is an important goal in ecological restoration, how is it conceptualised? To trace this question, let us go back to the specific example with which I began this section, that of restoring ecosystems in the post-mining landscapes of Lower Lusatia. With the help of this specific example, I will try to investigate what are considered as intact or functioning ecosystems in the practice of restoration. Nevertheless, given the high complexity of the example, I will be able to provide only a coarse analysis, which then serves to extend my considerations to other cases of ecological restoration and discussions within restoration ecology.

7.3.1 Restoration of post-mining sites in Lower Lusatia

Post-mining landscapes occur in many parts of the world. Open-cast mining (or strip mining), as applied frequently in lignite mining, is a very radical and large-scale 'use' or even 'consumption' of landscapes. Lower Lusatia, in the German state of Brandenburg, southeast of Berlin, is one of the largest mining areas in Germany. The first small-scale mining activities here started in the eighteenth century, becoming really large-scale, however, only with the development of large machinery during the early twentieth century (Haselhuhn and Leßmann, 2005). In 1998 about 75 000 ha of land had been converted into mine areas (Hüttl, 1998), and several villages had been 'removed' in the course of this expansion. For lignite strip mining, several metres of overburden (i.e. soil and sediment; in Lusatia 80–120 m) have to be removed to reach the coal, which is then exploited. This also involves a massive and deliberate lowering of the groundwater table, in some places up to 80 m or more (Hüttl, 1998). Huge holes in the ground are created, but as not all material is coal, non-coal materials (e.g. overburden) are deposited as dumps or heaps. What remains is a kind of 'moonscape', with craters and hills, creating completely new landscape forms and reliefs in the former plains of northern Lusatia, originally consisting mainly of wetlands, deciduous and mixed forest, agricultural areas, and – in historic times – human settlements (see Großer, 1995 for details).

The federal mining law of Germany (*Bundesberggesetz*) and more specific laws and regulations of its states ('*Länder*') mandate the conditions of reusability ('*Wiedernutzbarmachung*') of the mining sites, taking into account the interests of the public. Restoration, or '*Rekultivierung*', the most common term used in Germany (literally meaning 'recultivation', perhaps best translated as 'reclamation'), is a specific expression in this obligation. Specific plans are to be developed for which different administrative agencies and non-governmental interest groups, such as non-governmental environmental organisations, have to be involved. For the development of such reclamation/restoration plans, other laws besides mining laws are relevant, especially conservation law. The National German Conservation Act (*Bundesnaturschutzgesetz*, BNatSchG), for example, adds more specific (but still rather general) goals to the restoration of landscapes, emphasising a near-natural design ('*naturnahe Gestaltung*') of the impacted landscapes, the sustainable usability of natural goods, and the service capability [or efficiency] and functionality of the economy of nature ('*Leistungs – und Funktionsfähigkeit des Naturhaushalts*'). The latter expression is difficult to translate (I tried to provide an almost literal translation); it may with some justification be interpreted in terms of securing ecosystem functioning and – as I would add – ecosystem services. Beyond conservation interests are, of course, also the interests of agriculture, forestry, water use, tourism, and other economic and public sectors that play important roles in the development of the reclamation plans.

As everywhere, early restoration efforts were not guided much by scientific ideas, even less by the results of ecological studies in the areas (see Häge (1996) and Haselhuhn and Leßmann (2005) for the history of recultivation in Lusatia). More recent restoration efforts in the post-mining landscapes of Lusatia, after German unification in 1990, were at least accompanied (but still not guided!) by a variety of ecological studies, especially from the Technical University of Brandenburg in Cottbus (Hüttl, 2000; Wiegleb and Felinks, 2001; Tischew, 2004; Haselhuhn and Leßmann, 2005; Nixdorf et al., 2005; Wöllecke et al., 2007). The restoration of the East German mining areas was special within Germany because mining activities were stopped rather abruptly after German unification in 1990, and reclamation plans had to be developed in a very short period of time. The accompanying scientific investigations could only start in 1994–5, at a time when the development of many reclamation plans had already proceeded rather far. In the following, I will refer specifically to the publications of the scientists involved in the restoration projects.

Hüttl and Weber (2001, p. 322) express the overall goals of the restoration in terms of constructing ecosystems: 'The dumps are rehabilitated with the aim of constructing sustainable ecosystems.'

In another, more general paper on mine-site restoration, Bradshaw and Hüttl likewise argue that 'restoration depends on the restoration of whole ecosystems' (Bradshaw and Hüttl, 2001, p. 88) and urge for 'the creation of ecosystems that function properly in all respects and that are fully self-sustaining' (Bradshaw and Hüttl, 2001, p. 89).

As the post-mining area to be restored in the Lower Lusatia case is large, and the land forms resulting from the mining activities are diverse, several different types of ecosystems were (re)created. These types were selected in compliance with the regional planning requirements and the land use types predefined in the reclamation plans. Hüttl and Weber describe them as follows:

In the Lusatian mining district, there are four major ecosystem types re-established after mining: (1) forest ecosystems, mostly pine (Pinus sylvestris L.) or oak (Quercus spp.), predominantly on geogenic carbon-containing, 'black', acidic, sandy substrates; (2) agricultural ecosystems, preferably on more fertile soil substrates; (3) mining lakes, generally strongly acidic and initially considered 'dead'; and (4) naturally regenerating systems, on untreated 'succession sites'. (Hüttl and Weber, 2001, p. 322)

The fourth type of ecosystem was only included explicitly into the reclamation plans after insistent lobby work by conservation NGOs (Birgit Felinks, personal communication, September 2009).

214   ·   **Putting ecosystem functioning concepts into practice**

It was in fact possible to create a variety of different kinds of ecosystems in the area. After mining, most parts of the dumps and large pits were not 'predestined' to be inhabited by a specific 'natural' ecosystem type, at least not in the short run. The abiotic conditions had been severely altered, with the resulting new substrates being highly heterogeneous in structure and mineral content. Spontaneous succession in these areas with predominantly highly acidic substrates would take a long time to develop into, say, a forest ecosystem. So restoration was mostly actively 'assisting' the recovery (or better: re-creation) of ecosystems by methods such as amelioration of at least the upper soil layers, for example, by addition of limestone or other buffering substances. Also, geotechnical stabilisation of unstable slopes was often necessary. Nevertheless restoration could, in principle, be done in different ways and with different outcomes in terms of the future ecosystems to be (re)created.

So the specific ecosystems that were (re)created were based on decisions taken, which, for a specific area, could also have been taken differently. The reason for selecting them in the way they were selected was partly to mimic the pre-mining pattern of land use types, and also due to anticipated recreational uses. However, decisions were, of course, not independent of the climatic and historical biophysical setting. As indicated above, agricultural ecosystems were created predominantly on the more fertile soils (which were more promising for attaining the goal of provision of food or bioenergy production more rapidly); forest was created on less fertile soils. And even though the substrate on which the new ecosystems are about to develop was greatly altered with respect to historical conditions before mining, it is almost trivial to say that nobody (hopefully) would aim at creating a mangrove ecosystem in northern Germany, as there are limitations on the types of plants that can exist there under the available range of climatic conditions. In several cases, however, species not native to the region or even Europe were used for restoration (e.g. the oak species *Quercus rubra*, native to North America; Birgit Felinks, personal communication).

In principle, there are thus multiple options in creating functioning and 'sustainable' ecosystems for particular parts of the overall area. Beyond the general statements about this aim of (re)creating sustainable ecosystems, however, no further specific definitions of desired ecosystems and when they are considered to be properly functioning can be found. So, it is necessary again to interpret the texts with respect to this issue. This is what I will do in the following.

If we look at the four types of ecosystems described by Hüttl and Weber (2001), we find – at least at first – three, if not all four, of the ecosystem functioning types presented in Section 7.1. Forest ecosystems correspond largely to the 'physiognomic-type ecosystem', agricultural ecosystems to the 'process-focused type', here in its expression as an 'ecosystem-services type of ecosystem'. Succession sites correspond to a 'generic-type' ecosystem (here: 'wildness-type' ecosystem). Ideas on how mining lakes should look vary between different options for different lakes, from an almost generic-type approach to an ecosystem-services type of approach, to even a species-specific type of approach. It seems almost obvious that the criteria according to which these different systems will be assessed with respect to whether they are functioning ecosystems will clearly be different for each of them. But let's look closer at the different ecosystems.

Forest ecosystems
This ecosystem type is the most traditional, established restoration goal in the reclamation of mining land (Häge, 1996). The way it is established basically refers to a physiognomic-type ecosystem as sketched above. After prior improvement of the soil substrate through amelioration methods, the new forests are established by the planting of trees (here: *Pinus sylvestris*, *Pinus nigra*, *Quercus rubra* and other oaks, and *Robina pseudacacia*; most of them not native to the surrounding area), the dominant life forms of the target system. There has been no steering with respect to re-establishing specific undergrowth plant species or heterotrophic organisms. Herbaceous plants were seeded only in some experimental plots (Birgit Felinks, personal communication), but only with the explicit aim of erosion control (Hüttl, 2000, p. 56). The 'autonomous' succession of these other organism groups has, however, been the subject of scientific investigations to learn about the ecological processes going on in these developing ecosystems (Hüttl, 2000; Hüttl and Weber, 2001), as have been analyses of changes in abiotic factors and biogeochemical processes.

Forests are also created with respect to aesthetic aspects of the landscape, which, on the one hand, is a feature of the classical physiognomic idea, and on the other, may be seen also as a cultural benefit of these ecosystems. Forests are also created not only with respect to their physiognomy, but also with respect to selected *ecosystem services* in order to provide specific benefits: soil stabilisation, timber use, and recreation. They are somewhat intermediate between the types.

(a)

(b)

Fig. 7.3. Forest plantation in the post-mining area of Lower Lusatia. Photos: Courtesy of Birgit Felinks, August 1996 (a), October 1998 (b).

Agricultural ecosystems

Agricultural ecosystems have received the least attention in ecological studies during the restoration of the Lower Lusatian post-mining landscape. As discussed in Section 7.1.2, agricultural ecosystems are the paradigmatic type of an ecosystem-services type of ecosystem. Their proper functioning is given when the services they are designed for are delivered over a reasonable period of time. Long-term productivity here is a major indicator for the successful restoration of an agroecosystem. While these systems do fulfil the named criteria, their *degree* of functioning is in many parts still below that of comparable unmined areas in the region, i.e. they have a lower productivity. An extended description of research on the restoration of agricultural ecosystems in the area is given by Haubold-Rosar (2002, 2004).

Succession sites

Some areas were left to unaided succession, i.e. ecosystems were created where nature is allowed 'to take its course'. One might say that this is not restoration at all, because no *active* 'assistance' to the 'recovery of an ecosystem' (as expressed in the SER definition) is given. Many authors, however, also see the active and intentional refusal to intervene (also: *preventing* intervention) as a way of restoration (see Prach *et al.*, 2001; Prach and Hobbs, 2008, and critical discussion in Clewell and McDonald, 2009).

This specific type of target ecosystem is highly divergent from what was and is traditionally seen as a goal of restoration, both in the established practice of restoration and in restoration ecology. It was this type of landscape that I especially experienced as 'wild' more than a decade ago, areas that had been designated as priority areas for nature conservation. Although the original reclamation plans for the Lower Lusatian mining area envisioned that about 15% of the area would be designated for this purpose, less than 1% has actually been left to spontaneous succession (Wiegleb and Felinks, 2001). The sites have been subject to very intensive ecological research, using it – as with other parts of the post-mining landscape – as a fascinating experiment for research on ecosystem development. In the case of these succession sites, the underlying idea of a functioning ecosystem comes close to the generic definition of ecosystem functioning, and in a conservation context to what I have called the 'wildness-type ecosystem'. A functioning ecosystem here is not one that is characterised by a specific species composition, a specific physiognomy, and/or very specific processes or ecosystem services, but simply

(a)

(b)

Fig. 7.4. Different types of succession areas in the post-mining area of Lower Lusatia. Photos: Courtesy of Birgit Felinks, June 1997 (a), August 1995 (b).

by any ecological processes going on – with the specification, however, that these processes are not controlled by humans even though they have been initiated by human activities. As Wiegleb and Felinks (2001, p. 214) state: 'We regard the protection of natural dynamics as a high-ranking goal in nature conservation. Natural dynamics relates both to biotic (colonization, migration, succession), abiotic (soil dynamics, erosion, geomorphology) and combined processes (soil formation). Measure of initialization of soil and vegetation are regarded as unnecessary.'

Some types of processes are named, expected to occur, and even desired (especially those requiring large spatial and temporal scales (Felinks and Wiegleb, 1998)), but it is the property of having unfettered dynamics that characterises ecosystem functioning here.

The outcome of vegetation (and ecosystem) development is seen as uncertain and only very coarsely predictable (Wiegleb and Felinks, 2001), but also as not important for the goal of restoration to be achieved. For the succession areas in Lower Lusatia, Schulz and Wiegleb (2000, p. 107) explicitly state:

The developmental aims comprise the long-term guarantee to allow abiotic and biotic processes to take place in a sufficiently large undivided and undisturbed area. This also includes wind and water erosion. The definition of a desired vegetation composition and mosaic is not possible. All possible developments, from homogenous monodominant stands to a small-scale mosaic of vegetation types, are allowed and will be accepted.

Wiegleb and Felinks (2001) perceive of the succession areas as being in a 'near to natural state' due to this free-running dynamic. In the light of the high ambiguity of the term 'natural', a better expression for what is meant here would be 'wildness' (see Section 7.1.1 above). In the highly managed landscapes of Germany, wildness in this sense (not to be confused with 'being in a *pristine* condition') is highly valued by many conservationists, expressed in the strategy that has been called *Prozessschutz* (process conservation) in Germany (Sturm, 1993; Felinks and Wiegleb, 1998; Potthast, 2000b; but see my remarks in Box 7.1).

Lakes

Around 25% of the earlier mining areas will develop into open water bodies (Schulz and Wiegleb, 2000). A large number of lakes originate from the mining holes as a result of the mass deficit brought about by the extracted lignite, and ground and surface water filling in after mining activities have ceased. The lakes are very diverse with respect to their morphology; some of them are very deep and large. The definition of goals for this kind of ecosystem is complicated. Such lakes are a new landscape element in Lower Lusatia, one that was not present in the history of the area. In fact, creating and developing lake ecosystems in this area is not really *restoration* any more, but creating a novel type of ecosystem for this region. Most of the new lakes start from extremely acidic conditions and, even though they are not 'dead', they are colonised only by a few species of microorganisms (bacteria, protists, some zooplankton) at first (e.g. Nixdorf *et al.*, 1998; Wollmann *et al.*, 2000). Nevertheless, restoration goals have to be formulated for these new systems. These goals are strongly determined by the requirements of the WFD, discussed in Section 6.2, which is also relevant for impoundments and other artificial lakes. The management goal for these types of waters is not 'good ecological status', but 'good ecological potential'. That is, the presumed 'natural state' of the ecosystem is not the management target, but the ecological state which can be reached at best by means of management, given the specific abiotic setting. The ultimate reference point here is the 'maximum ecological potential' as those quality elements (see Section 6.2 above for WFD terminology) that reflect the 'closest comparable surface water body type, given by the conditions which result from the artificial or heavily modified characteristics of the water body' (European Community, 2000, Annex V, table 1.2.5). 'Good ecological potential' is defined by only a slight deviation of, say, the biological quality elements. The WFD leaves a lot of room here, and several ideas have been discussed by the scientists involved. Should the lakes (at least some of them) be kept in the specific acidic state, left to natural succession in the sense of *Prozessschutz* (Leßmann and Nixdorf, 2002)? As there are no direct reference systems (naturally highly acidic lakes) for most of the mining lakes available in the direct vicinity and even in the biogeographic region, should reference systems be sought in other parts of the world? For most lake types occurring in the post-mining areas of Lusatia, Nixdorf *et al.* (2005) identified natural lakes with comparable *abiotic* conditions in far away places, such as acidic volcanic crater lakes or coastal sulphur lakes found in Argentina or Japan. For some of the mining lakes, however, it was not possible to find any naturally occurring reference lakes at all (Nixdorf *et al.*, 2005). Another way of defining reference conditions in the sense of good ecological potential would be to select a mesotrophic, pH-neutral lake to be used for bathing, other recreational activities, and fisheries, and to look for reference lakes conforming to these conditions (Leßmann and Nixdorf, 2002). To achieve this in a reasonable period of

Fig. 7.5. New lake forming in a former mine pit in Lower Lusatia. Photo: Courtesy of Birgit Felinks, August 1996.

time would, of course, require the intense use of remediation measures, such as external flooding by alkaline water, chemical neutralisation, and other measures (Nixdorf *et al.*, 2005). While most of the lakes might be remediated for direct human use, there are postulates by the scientists to preserve some 'selected very and extremely acidic lakes [. . .] as valuable and extreme and unique ecosystems' (Nixdorf *et al.*, 2005, p. 71).

So there are several ideas of a functioning ecosystem here, which may be pursued differently for different lakes: (1) an approach coming close to a generic ecosystem approach – in the bounds of specific abiotic factors – where nature takes its course; (2) that of the ecosystem-services type of ecosystem for human use; and in principle – following the ecosystem integrity approach underlying the WFD – (3) even a species-specific type of ecosystem functioning. How to compare the ecological state of specific mining lakes with the selected biological reference conditions (as decisive for assessing whether good ecological potential is reached) still remains somewhat elusive in the literature, especially when it comes to reference lakes on other continents, with a completely different biogeography.

Post-mining landscapes, in comparison to the case studies from Navarino Island (Chapter 2) and Yellowstone National Park (Chapter 4), are certainly at the other end of a gradient of ecological systems, the latter almost not modified by humans and the former completely dominated (or even destroyed) by them. While there may still be rather good historical reference states in the Yellowstone and Navarino cases, reference to such states seems inappropriate for the heavily modified landscapes of post-mining areas. But even ecosystem functioning types described in the Lusatia example are not completely devoid of comparisons to historical states of the former ecosystems. Even though the main aim here is to gain insight into how ecosystems 'work', Hüttl and Weber (2001) compare their results on chemistry, biogeochemistry, and biology to the conditions on nearby unmined areas. The purpose, however, is not to really *mimic* these conditions on the restoration sites, as such an aim would not be realistic in practical terms.

The example demonstrates also that the four ecosystem functioning types I discussed in Section 7.1 are not strictly delimited types, but that many kinds of transition or intermediate conceptualisations of functioning ecosystems exist.

Although the construction of functioning ecosystems is expressed as an explicit restoration goal by some researchers, a more precise explicit *definition* of the desired ecosystems is not given, nor is a description of the criteria they consider 'functioning' to cover. This information could be inferred only indirectly, as that about specific reference states. The selection of the target ecosystem on such a large scale has been a process that was first constrained to a considerable degree by legal requirements as, for example, given by mining law or the WFD. Beyond that, decisions for developing reclamation plans were taken at the level of the state of Brandenburg in a committee that consisted of representatives of elected political institutions and selected interest groups, among them environmental NGOs, unions, farmers associations, and chambers of commerce. The resulting share of different areas and the ecosystems to be restored on them constitutes a compromise between the different interests – within the boundaries given by the legal and the biophysical situation. Only for the conservation areas (and to a minor degree for the lake areas; see Leßmann and Nixdorf, 2002) has there been a systematic, detailed scientific search for specific goals. Besides what was selected in the end, the aim could have been, for example, reconstruction of some endangered habitat types, restoration of the traditional cultural landscape, maximisation of biodiversity, maximisation of habitat for endangered species, etc. (the latter two goals have in fact been pursued in some areas designated for conservation purposes; Birgit Felinks, personal communication). The idea, then, was to initiate a societal discourse on *Leitbilder* ('guiding principles', or 'target visions'), at least for nature conservation in the post-mining areas (Blumrich et al., 1998; Felinks and Wiegleb, 1998; Schulz and Wiegleb, 2000; Wiegleb et al., 2000). These guiding principles involved clarification of a large array of choices with respect to possible and desired ecological states and societal values on different spatial levels. The discussions thus did not explicitly focus on specific 'ecosystem' goals of conservation.

To my knowledge, the very thoughtful and innovative ideas and procedures developed by the researchers in the context of conservation-related restoration of post-mining areas have not found broader implementation in the planning process of Lusatia; this is a general problem of the interaction between science and policy. The ideas developed in the *Leitbild* discourse have, however, at least locally and punctually, led to consideration of alternative strategies of restoration to those used traditionally (B. Felinks, personal communication, September 2009). This led, for example, to a continuation of '*Prozessschutz*' (unaided succession) in some restored areas after they changed from the ownership of the mining company to others; in a few cases purely physiognomic – and ecosystem service-centred – views of forest ecosystems were refined to that of ecosystems whose components (here: species) were more specifically defined (increasing the number of native tree species).

7.3.2 Beyond Lusatia: ecological restoration and ecosystem functioning

As I said above, ecological restoration has many goals. Although preserving the (proper) functioning of ecosystems is not the only goal, it is increasing in importance, as the SER's definition of ecological restoration ecology, cited above, implies. As in the application of BEF research, it is often not completely obvious if the postulates to preserve ecosystem functions refer to selected processes or to the overall functioning of the system. If the latter is meant, its precise meaning mostly remains somewhat unclear. The expression 'healthy ecosystem' is often used, as is the phrase 'ecosystem integrity'; both phrases are rejected by other authors for the reasons discussed above (Section 6.2).

It is not possible to provide an extensive analysis of concepts about ecosystem functioning in restoration projects beyond our specific example from Lusatia. One 'classical' idea in restoration *ecology* (not necessarily the practice of restoration in the broader sense) is certainly what I have called the 'species-specific type' of ecosystem functioning. It is especially promoted where ecosystems are seen as only slightly degraded, such as within national parks like Yellowstone (Section 4.3). The pristine or 'true wilderness' areas are most likely to be *restored* by such an approach. The idea of almost every detail of nature being in its 'natural' or historic state is still a rather prominent idea of what an intact, functioning ecosystem is, as demonstrated by the SER criteria for successful restoration. As might not be surprising, this idea is contested (e.g. Hobbs and Norton, 1996; Choi, 2007; Dufour and Piégay, 2009) and involves choices, beginning with choosing the appropriate reference date in history (Higgs, 2003). I have already discussed the problem of what a 'natural' ecosystem could be in the context of the Yellowstone case study; it applies in the same manner here.

Underlying the goal of looking at recent reference systems (rather) unimpacted by human activities, or at historical systems of the same kind is the assumption that these systems are functioning properly, i.e. that they are stable and/or resilient, and even have the potential to provide many desired products and processes (ecosystem services) which results in benefits to humans. In a way, it is the assumption that 'nature knows best'. Given our ignorance of many ecological processes and their outcomes, it may not be a bad assumption, while the idea that we are able to design and steer ecological systems perfectly would certainly be hubris. Nevertheless, these systems alone will not be able to provide all the services on which more than five billion people depend (e.g. to provide enough food) when running without human intervention. Also, climate change and other global changes will alter these systems at a much faster pace than before. So the task is to understand the processes better and to carefully tinker with the aid of those ecological principles which we already understand.

In terms of my coarse classification of ecosystem functioning types, the second important understanding of a properly functioning ecosystem as a restoration target refers to the 'process-focused type', specifically its subtype, 'ecosystem-services type'. This is by far not restricted to purely agricultural areas. Restoration success is increasingly understood in terms of ecosystem functioning as the delivery of desired ecosystem services and benefits, up to the design of completely new ('novel') ecosystems. Thus Palmer *et al.* (2004, p. 1252) state: 'However, "designing" ecosystems goes beyond restoring a system to a past state, which may or may not be possible. It suggests creating a well functioning community of organisms that optimize the ecological services available from coupled natural-human ecosystems.'

Others join this call for new directions in restoration (Dufour and Piégay, 2009; Choi, 2007). Choi, who even demands a new 'paradigm' for restoration ecology, writes: 'Future-oriented restoration should focus on ecosystem functions rather than recomposition of species or the cosmetics of the landscape surface' (Choi, 2007, p. 352).

The boundary between restoration and new design of ecosystems becomes blurred here, which is, however, a logical outcome of the observation that we can never arrive at a 'carbon copy' of specific physical reference ecosystems (Higgs, 2003; Hilderbrand et al., 2005). Nevertheless, even with a focus on ecosystem services, Aronson et al. (2007, p. 7) suggest that 'self-sufficient, self-organizing, *natural* ecosystems' (emphasis added) are most likely to provide most ecosystem services better than human-designed systems.

The different ecosystem types and the different conceptions of ecosystems mirror different purposes underlying restoration efforts as described above. They thus show how even implicit decisions have a strong influence on praxis. But how open and clear are these decisions, by which motivations are they really guided? And, especially, who makes the choices and who takes the decisions to put things into practice?

Who makes the choices?
Although there is a lot of discussion on goals of ecological restoration, and also on the values and cultural aspects underlying it within the scientific literature, in the practice of restoration, rather little consideration is given to such contexts (Higgs, 1997) and even to the very rationales of restoration, as Clewell and Aronson (2006, p. 421) state: 'Descriptions of restoration projects frequently ignore the why of the project and imply that the need for restoration is inherently obvious and its intentions are noble. The underlying reasons to restore remain understated and unappreciated.'

The fact that definitions of ecosystems and ecosystem functioning are not given by nature but are partly a matter of choice opens up one road for societal decisions in restoration ecology – for good and for bad. For good, it means that participative and democratic processes, aspects of culture and ethics have a very clear interface with restoration efforts. The restoration of ecosystems, be it for the direct benefits and wellbeing of people or for that of the other organisms that live in it, is an important societal issue that should not just be a matter of technocratic procedures guided by politicians and 'experts'. The flip side of ecosystem functioning being a matter of societal choices is that it might be taken as a carte blanche for declaring *any* state of ecological systems as equally valid and 'functioning'. This is what the opponents of constructivistic approaches to nature fear most. I think that the danger of unclear goals and of hidden assumptions leading to undesired, unanticipated, and unsupported outcomes of restoration efforts (and other measures of conservation in the broadest sense) is much greater than the danger of opening up a discussion. I have dealt with this issue above (Section 5.1) and I will come back to it briefly in the last part of the book (Section 7.4).

Restoration ecologists largely agree that restoration requires choices and is dependent on societal values (e.g. Hobbs *et al.*, 2004; Allison, 2007, the former comprising a selection of several short papers from different authors). In the literature about restoration ecology there is a discussion on how to involve stakeholders on various spatial and administrative scales; various examples exist where this has been done in practice (e.g. Allison, 2007; Burke and Mitchell, 2007; Junker *et al.*, 2007). Smaller, local projects are easier to deal with in such a way than large ones affecting whole regions, such as that of restoring post-mining landscapes in Lusatia or other parts of the world. Details about the issue of including different interest groups in the practice of ecological restoration can be found in the papers cited above and in the books of Gobster and Hull (2000), Gross (2003), or Higgs (2003).

Restoration goals can be expressed in terms of ecosystem functioning, but they do not have to be. If we agree, however, that restoration should refer to the larger system, i.e. the ecosystem, the idea of formulating it in terms of ecosystem functioning can be highly useful.

7.4 Conclusions and outlook

Ecosystem functioning, as dealt with in this book, is the overall performance and working of an ecosystem. In its generic form it denotes any processes going on in a system composed of organisms in interaction with their environment (being the generic ecosystem definition; see Chapter 4). There is thus no way in which such an ecosystem can *not* be functioning. In more narrow meanings of 'ecosystem functioning', however, in which the phrase is used most commonly, it denotes *proper* functioning, i.e. *the continued performance of a system within some limits specified as reference conditions (state, dynamics or trajectory)*. In such specific meanings, assessing ecosystem functioning depends on the specific ecosystem definition embraced, a specific reference condition, and specific spatial and temporal scales of observation. Ecosystems, then, may also be considered to be not properly functioning, or not functioning at all. Given the wide possibilities for defining ecosystems and their reference conditions, no universal idea of a functioning ecosystem is possible, but only relative ones – relative to specific definitions and the purposes they are used for. Making statements about ecosystem functioning precise and measurable in consequence involves a number of methodological and/or social choices that have to be made – as described in Section 7.2. Defining and assessing ecosystem functioning thus involves both scientific and normative dimensions.

This diagnosis, which I have developed throughout this book, and the parts of which I have illustrated by a number of case studies, has several implications for ecology and its application fields. First, it necessitates rethinking and even restructuring parts of the research on ecosystem functioning. Second, it has implications for the role of the researcher in assessing ecosystem functioning, especially in applied contexts such as conservation biology, ecosystem management, and restoration ecology. It puts interdisciplinarity, societal concerns, and ethical issues high on the agenda. I will discuss these implications of my analysis briefly in the remainder of this book.

7.4.1 Ecosystem functioning as a research programme?

This book has dealt with how to conceptualise and assess ecosystem functioning. It has, however, not – or only at some points in passing – discussed which *mechanisms* are responsible for maintaining ecosystem functioning. This would be the task of another book. A prerequisite for this task, however, is to be more explicit about what ecosystem functioning really can mean, as elaborated here.

Given the breadth of possibilities for conceptualising ecosystem functioning in a meaningful way, we should not attempt to look for a general theory of ecosystem functioning or even for a general theory of BEF relationships. On the other hand, not every definition of an ecosystem and of ecosystem functioning must and should be unique. What we should strive for is to subdivide the universe of possible concepts of ecosystem functioning into more restricted domains for which useful and predictive theories of *restricted* generality can be formulated. This would be a step towards (re)structuring a research programme on ecosystem functioning. Types such as those described in Section 7.1 might be a starting point, but as could be seen in the case study on ecological restoration above (Section 7.3), there may also be other, intermediate ones. Characterising useful types of ecosystem concepts and related concepts of ecosystem functioning is a matter of choices, conceptual work, and empirical work.

We may aim at some ideas of a functioning ecosystem that are not feasible, or not feasible any more. Given the fact of climate change, for example, some ideas of 'self-sustained' functioning ecosystems as characterised by a particular (historic) species composition might become possible in the future only at the expense of continued and active human intervention, or even become impossible. Instead, we will have to expect many novel ecosystems ('no-analog systems') in the future (Fox, 2007; Hobbs *et al.*, 2009). These are not conceptual matters or choices, but a matter of the ecological conditions under which species can thrive or not thrive − independent of our choices and concepts. This does not mean that we have to turn in such cases to either an ecosystem-services idea of functioning or even a generic idea of functioning. We may still try to maintain some of the parts (species), depending on the actual possibilities given by nature.

I am convinced that it is worth the effort to restructure research on ecosystem functioning according to more refined definitions of the concept. The four ideas presented above are certainly too coarse in this respect. The process–focused ecosystem functioning type (Section 7.1.2), in particular, still allows an extremely broad range of possible ecosystem definitions which should be further subdivided. A good candidate type to start with may be the physiognomic type of ecosystem functioning (Section 7.1.3), not the least as it is often connected to thresholds and correlated to a specific set of processes.

Much of the task of structuring the universe of ecosystems and ecosystem functioning concepts can be done without relations to societal concerns. But if it comes to determining ecosystem functioning as the target (or part of the target) of conservation, management or restoration, societal choices become indispensable. They may, in turn, even facilitate a useful selection of meaningful ecosystem functioning types. Selecting ecosystem functioning for applied purposes has to be focused on values and only then on the resulting target systems. The ethical dimension is paramount in deciding about what a *desired* functioning ecosystem is. It amounts to nothing less than how we want to live in our world, what counts as a good life for us.

7.4.2 Is ecosystem functioning good?

It may be almost superfluous to say that ecosystem functioning is not automatically something good and desirable. There may be highly undesirable states of ecosystems which nevertheless, according to some definitions of ecosystem functioning, are performing perfectly (e.g. a hypertrophic lake ecosystem). The question of whether a particular state and/or mode of functioning is preferable to another is beyond any natural science, it is a matter of societal discourses and ethics. Different systems may exist and 'function' on the same site: a meadow, forest, or cornfield, etc. All of them may be considered as functioning ecosystems, but on behalf of different criteria. Also, some functioning ecosystems (e.g. natural wetlands, deserts) have been considered as highly undesirable by humans and converted into other types of ecosystem. Opting for one type of ecosystem functioning as a goal of management often excludes other possible types: an ecosystem-services type of ecosystem functioning may contradict a species-specific type of ecosystem functioning, especially when the former is realised through an agricultural ecosystem. There are trade-offs, and these trade-offs are morally relevant: for the benefits which different groups of people derive from the ecosystems, for other species involved in the ecosystems. Choices about protecting or restoring ecosystems are, fundamentally, also about ethics. What is worth protecting? What do we value? Who values what? How do we want to live? Do not make the mistake of thinking that such questions are just to be answered as a matter of moods and subjective preferences. Ethics, as a discipline of philosophy, is based on sound and rational arguments, which must be made explicit and open to discourse. It is also about what we consider as a good life, about our relations and obligations to other humans, and – more recently – about our relations and obligations towards the non-human world.

Only a minority of philosophers views ecosystems as 'moral objects', with respect to which direct moral duties may be formulated (but see Rolston, 1994; Gorke, 2007). This issue is further complicated by the very ambiguity of the ecosystem concept. In fact, we rarely value ecosystems per se. Ecosystems are almost always a *means* for something. It is some objects (individuals, species, species types), processes and their products (including the aesthetics of particular physiognomic appearances) which we aim at – in their respective contexts. Balancing different values attributed to ecological systems is a challenging issue, even within the same moral system. And there is not *the* one ethics of the environment. Environmental ethics is a highly dynamic field of philosophy, with many different approaches, using different criteria for moral relevance of objects and weighing them against each other in different ways (for overviews see Birnbacher, 1980; Hargrove, 1989; Hampicke, 1993; Eser and Potthast, 1999; Schmidtz and Willott, 2002; Desjardins, 2006).

Likewise, different groups of humans may judge differently whether the functioning of an ecosystem is good for them. The classical conflicts between biodiversity conservation and indigenous people are an example here: a 'pristine' functioning forest ecosystem (defined via its physiognomy and the services it delivers to further climate regulation) may be valued as good by conservationists and even by many governments caring about climate change, while it may be valued as useless or even negative by local people who want to convert it into an agroecosystem.

The case of the beaver on Navarino Island as described in Chapter 2 is another example of how different methodological choices and different value systems affect whether changes in an ecosystem are seen as impairing functioning, and if this impairment is a problem that requires action.

As the research of Anderson and colleagues (Anderson *et al.*, 2006b, 2009; Anderson and Rosemond, 2007) showed, the beaver has altered many patterns and processes on Navarino Island. On the reach scale and the stream scale it changed species compositions, fluxes, and physiognomy, i.e. with respect to many different definitions, ecosystem functioning was altered, or ecosystems were even transformed into others (from fast-flowing to almost-standing waters). The question is: is it a problem? This depends exclusively on the values attributed to particular states and types of ecosystem. The new (small-scale) ecosystems created by the beavers' activities may likewise be considered to be properly functioning (as a beaver-pond ecosystem, for example), depending on the reference states defined. Also, on a larger scale, no species is known to have become extinct on the island, and so we may consider the whole island to be functioning properly under the perspective of a species-specific ecosystem definition (unless we see the addition of species as a significant impairment of ecosystem functioning). Beyond that, we may consider ecosystems as always changing, with the appearance of new species as altering but not impairing ecosystem functioning. It is thus neither evident that the beaver really impairs ecosystem functioning on Navarino Island (as this judgement depends on the definition embraced) nor is it evident which action should be taken with respect to the beaver. The latter is completely a matter of value decisions (see Haider and Jax, 2007 for a discussion of this issue in the light of different ethical approaches). It cannot be derived from scientific work alone, even though it must clearly be informed by empirical facts. There is thus (and can be) no scientific reason to eradicate the beaver on Navarino as is currently being attempted (Choi, 2008). Clarification of the different ideas of ecosystem functioning embraced by the scientists (and of alternative possible definitions of ecosystem functioning with respect to specific settings), as well as the reasons *why* these specific definitions have been chosen, can help to open the discussion on the possible consequences of the scientific studies to scientific and societal discourse. This requires, however, a change in how scientists view their own role in a conservation context.

7.4.3 The role(s) of scientists in ecosystem functioning research (and beyond)

If conceptualising and assessing ecosystem functioning is a task that involves normative choices, then it is not only the role of society at large that changes with respect to research on this topic. It is also the role of that part of society that professionally acts out its curiosity that changes: the scientist themselves. Science has changed dramatically during the last century. While science has never really been de-coupled from the other parts of society, scientists – and ecologists, conservation biologists, and other environmental scientists prominent among them – have increasingly been drawn into the public arena, into policy issues. With these new challenges, several different scientist roles are possible and needed (see Pielcke, 2007 for a very intriguing analysis). Scientists may try to step back and find a neutral role, but often this is not possible. When scientists conduct research on ecosystem functioning they have already made decisions about specific ideas as to what a functioning ecosystem is for them, and have thus excluded other ideas. This is unavoidable and not a problem as such – as long as there is consciousness that other possibilities for conceptualising ecosystem functioning exist, and as long as this is communicated when the results of research are used for practical purposes, such as conservation policies. Otherwise, the scientists become advocates, openly or – and here comes the real problem – in disguise (a 'stealth issue advocate' as Roger Pielcke (2007) has called it). There is no real problem in being both an advocate and a scientist. In this case, however, the different hats of the person should be visible; scientific authority should not be used as disguised advocacy. Given this, the changing role of the scientist means that he or she must be more aware of his or her own values and choices – and how they guide actions at every stage of the research process. It also implies an increased necessity or even duty for communicating research. And communication here should not only travel the traditional one-way road of 'informing' society, but should also start a true dialogue, in many cases involving an ambitious translation process (see Section 7.2). Such demands easily overburden many scientists, which adds a further argument for the need of interdisciplinary studies in conservation biology, bringing together researchers from the sciences and the humanities.

The challenge is to bring science and society together, but to avoid classical – or one might say: traditional – pitfalls. It involves the relation of ecologists and conservation biologists to social scientists (interdisciplinarity), as well as their relation to society at large (transdisciplinarity). Both kinds of relations are often mixed up, but have to be dealt with as distinct issues. A standard practice in conservation biology has been to first do the natural science and afterwards bring in some social scientists, whose task is then seen as either communicating the scientific results (and the decisions implied in them) or do research on their acceptance. Social science research is then seen more as a formal add-on to the 'real stuff'. The role of the natural scientist with respect to society is then that of the advocate, using the reputation of their field as 'objective' research – no matter if this is a conscious act or not. The reason for this may be a high commitment to some environmental issues or it may be a naive understanding of ecology as a value-free science in which the 'facts' discovered by science are so compelling that they render any value decisions completely superfluous. On the other hand, social scientists – if they care about conservation issues at all – tend to consider environmental problems almost exclusively as social problems and often either work with very simplified ecological ideas and/or do not care about the results of scientific research at all. While this may be an exaggeration to some degree, it is not so far from what I have experienced in many cases. Cooperation between ecologists and conservation biologists on the one side and social scientists – and even theoretically minded ecologists that venture into the more philosophical side of theory – is still a difficult task. It is hampered by missing knowledge about each other's methodologies and, connected to that, biases against the other disciplines. There are, of course, also 'technical' problems in bringing together the different methodological approaches.

To foster interdisciplinarity we must strive for curricula that introduce students of ecology and conservation biology early in their career to basic ideas of methodologies from the humanities, such as epistemology, ethics, hermeneutics, and the history of concepts. The form and contents of education as it has been practised mostly within the different scientific disciplines leads to the rapid and lasting hardening of a specific code of traditionally customary methodologies and ways of thinking within each discipline. As a consequence, methodologies from other disciplines – especially if they are from more distant fields – are mostly not known at all or at least are viewed very sceptically. Every disciplinary stock of knowledge and methods is also connected to specific theoretical and philosophical assumptions about reality and about the adequate approaches for its investigation. Because of this it often becomes very difficult – especially between the natural sciences and the humanities – to develop an understanding of the work of the other disciplines, which in part leads to the complete disapproval of other approaches. Each discipline's own way of gaining knowledge is perceived as the only way that makes sense. This frequently leads to a failure of interdisciplinary communication, but also to a failure of a self-critical reflection of one's own fundamentals (here: those of ecology). This, however, is of high importance for the further development of the still-shaky theory-building of ecology.

A basic knowledge of these different methodological approaches – taught with relation to ecology and conservation biology – will provide an understanding of the validity and potential of the different thought-style of the humanities. It would greatly facilitate interdisciplinary studies and transdisciplinary approaches.

7.4.4 How useful is the ecosystem functioning concept?

Given its ambiguity and the difficulties in operationalising ecosystem functioning, is it a useful concept at all? What is the advantage of using an ecosystem functioning parlance?

There is a clear demand for overarching, integrating concepts describing environmental quality. Even though referring to the state and functioning of ecosystems is certainly not a panacea for all kinds of conservation and management goals, it has many fields where it can be applied meaningfully, as demonstrated in this book. Many goals of ecological conservation, restoration, and resource management can also be expressed differently, without referring to ecosystem functioning at all. But ecosystem functioning in its diverse expressions (also as ecosystem integrity, resilience, etc.) adds another perspective that draws together many otherwise unconnected variables of ecological systems, emphasising the internal dynamics of ecological systems and the maintenance of a particular state or dynamics. The problems of the concept, its ambiguity and its dependence on (social) choices are at the same time its strengths. The ambiguity allows the concept to be used as a boundary object (like biodiversity, resilience, and other conservation-related concepts), which has different meanings to different people, and to which most people, at first sight, can agree as a goal of management, and which thus brings people together. To be applied in specific circumstances (and also to derive scientific generalisations), however, we must become precise. As I tried to demonstrate, the concept *can* be made more precise for specific purposes. Only then can ecosystem functioning be a useful common denominator and focus for conservation, restoration, or management goals, opening up discourses and not hiding nebulous ideas and power relations behind the seemingly common goal of maintaining or creating functioning ecosystems.

If we acknowledge that assessing 'ecosystem functioning' and even more, defining what a 'functioning ecosystem' is, involves both scientific and social dimensions, the concept becomes open, is open to societal discourse. It does this without compromising scientific rigour – as long as we clearly distinguish between the different dimensions.

'Ecosystem functioning' can become a buzzword. It often is *just* a buzzword. If it only remains as such, we should get rid of it as soon as possible. But the concept, used consciously, can also be a powerful tool for analysing the state of our environment. It is relative to our definitions, but it is not relative to physical reality and to the many and severe influences we exert on the living world. Species *do* go extinct, climate *is* changing, human land use *is* increasing rapidly, artificial substances *are* increasingly released into the environment. There is no doubt about all these things. There is no doubt that they affect ecosystems and sometimes change them radically. There is no doubt about all this and there is still much scientific knowledge necessary to understand how ecosystems – in the different ways we can conceptualise them – work and how exactly they change. Having to choose between different ideas of what properly functioning ecosystems are does not leave the fate of the living world to relativity, as it does not deny that ecosystems change rapidly under human influence. To the contrary: it increases our responsibility for our fellow living beings, both human and non-human. Our concepts of ecosystem functioning do not remain mere descriptions, they have moral consequences: for ourselves, for other humans, and for the other organisms on our wonderful planet. Describing how ecosystems work, i.e. which processes, which interactions occur, is mostly a matter of science. Defining and assessing what a functioning ecosystem is, what proper functioning means, is as much about ecology as it is about what we consider as good life, and how we value the good lives of other people and other species. Precise and consciously applied concepts of ecosystem functioning can help to better assess the consequences of our activities and to formulate aims for conservation and natural resource management – connecting science and society.

References

Abele, L. G., Simberloff, D. S., Strong, D. R. J., and Thistle, A. B. (1984). Preface. In *Ecological Communities: Conceptual Issues and the Evidence*, ed. Strong, D. R. J., Simberloff, D. S., Abele, L. G., and Thistle, A. B. Princeton, NJ: Princeton University Press, pp. vii–x.

Adger, W. N. (2006). Vulnerability. *Global Environmental Change*, 16, 268–81.

Ahl, V. and Allen, T. F. H. (1996). *Hierarchy Theory. A Vision, Vocabulary, and Epistemology*, New York, NY: Columbia University Press.

Allen, T. F. H. and Hoekstra, T. W. (1992). *Toward a Unified Ecology*, New York, NY: Columbia University Press.

Allen, T. F. H. and Starr, T. B. (1982). *Hierarchy: Perspectives for Ecological Complexity*, Chicago, IL: University of Chicago Press.

Allison, S. K. (2007). You can't not choose: embracing the role of choice in ecological restoration. *Restoration Ecology*, 15, 601–5.

Andersen, T., Carstensen, J., Hernández-García, E., and Duarte, C. M. (2009). Ecological thresholds and regime shifts: approaches to identification. *Trends in Ecology & Evolution*, 24, 49–57.

Anderson, C. B. and Rozzi, R. (2000). Bird assemblages in the southernmost forests in the world: methodological variations for determining species composition. *Anales del Institutio de la Patagonia*, 28, 89–100.

Anderson, C. B., Rozzi, R., Torres-Mura, J. C., Mcgehee, S. M., Sherriffs, M. F., Schuettler, E., and Rosemond, A. D. (2006a). Exotic vertebrate fauna in the remote and pristine sub-antarctic Cape Horn Archipelago, Chile. *Biodiversity and Conservation*, 15, 3295–313.

Anderson, C. B., Griffith, C. R., Rosemond, A. D., Rozzi, R., and Dollenz, O. (2006b). The effects of invasive North American beavers on riparian plant communities in Cape Horn, Chile: Do exotic beavers engineer differently in sub-Antarctic ecosystems? *Biological Conservation*, 128, 467–74.

Anderson, C. B., Martínez Pastur, G., Lencinas, M. V., Wallem, P. K., Moorman, M. C., and Rosemond, A. D. (2009). Do introduced beavers *Castor canadensis* engineer differently in South America? An overview with implications for restoration. *Mammal Review*, 39, 33–52.

Anderson, C. B. and Rosemond, A. D. (2007). Ecosystem engineering by invasive exotic beavers reduces in-stream diversity and enhances ecosystem function in Cape Horn, Chile. *Oecologia*, 154, 141–53.

Angermeier, P. L. and Karr, J. R. (1994). Biological integrity versus biological diversity as policy directives. *BioScience*, 44, 690–7.

Aronson, J., Floret, C., Le Floc'H, E., Ovalle, C., and Pontanier, R. (1993). Restoration and rehabilitation of degraded ecosystems in arid and semi-arid lands: I. A view from the south. *Restoration Ecology*, 1, 8–17.

Aronson, J., Dhillion, S., and Le Floc'H, E. (1995). On the need to select an ecosystem of reference, however imperfect: a reply to Pickett and Parker. *Restoration Ecology*, 3, 1–3.

Aronson, J., Milton, S. J., and Blignaut, J. N. (2007). Restoring natural capital: definitions and rationale. In *Restoring Natural Capital: Science, Business, and Practice*, ed. Aronson, J., Milton, S. J., and Blignaut, J. N. Washington, DC: Island Press, pp. 3–8.

Ashby, W. C. (1987). Forests. In *Restoration Ecology*, ed. Jordan, W. R. I., Gilpin, M. E., and Aber, J. D. New York, NY: Cambridge University Press, pp. 89–108.

Attiwill, P. M. (1994). The disturbance of forest ecosystems: the ecological basis for conservative management. *Forest Ecology and Management*, 63, 247–300.

Baerlocher, F. (1990). The Gaia hypothesis: a fruitful fallacy? *Experientia*, 46, 232–8.

Bailey, R. G. (1996). *Ecosystem Geography*, New York, NY: Springer.

Baker, B. and Hill, E. P. (2003). Beaver (*Castor canadensis*). In *Wild Mammals of North America: Biology, Management, and Conservation*, ed. Feldhamer, G. A., Thompson, B. C., and Chapman, J. A. (2nd edn) Baltimore, MD: Johns Hopkins University Press, pp. 288–310.

Ballentine, R. K. and Guarraia, L. J. (eds) (1977). *The Integrity of Waters: Proceedings of a Symposium, 10–12 March 1975*, Washington, DC: US Environmental Protection Agency.

Balvanera, P., Kremen, C., and Martinez-Ramos, M. (2005). Applying community structure analysis to ecosystem function: examples from pollination and carbon storage. *Ecological Applications*, 15, 360–75.

Balvanera, P., Pfisterer, A. B., Buchmann, N., He, J.-S., Nakashizuka, T., Raffaelli, D., and Schmid, B. (2006). Quantifying the evidence for biodiversity effects on ecosystem functioning and services. *Ecology Letters*, 9, 1146–56.

Barbee, R. D. (1968). A discussion of ecological management in the national park system. MS thesis, Colorado State University.

Barkai, A. and McQuaid, C. (1988). Predator–prey role reversal in a marine benthic ecosystem. *Science*, 242, 62–4.

Barry, D. and Oelschlaeger, M. (1995). A science for survival: values and conservation biology. *Conservation Biology*, 10, 905–11.

Bastian, O. and Schreiber, K.-F. (eds) (1994). *Analyse und ökologische Bewertung der Landschaft*, Jena: Gustav Fischer.

Becker, E. and Jahn, T. (2006). *Soziale Ökologie: Gründzüge einer Wissenschaft von den gesellschaftlichen Naturverhältnissen*, Frankfurt/New York, NY: Campus.

Beierkuhnlein, C. and Neßhöver, C. (2006). Biodiversity experiments: artificial constructions or heuristic tools? *Progress in Botany*, 67, 486–535.

Beisner, B. E., Haydon, D. T., and Cuddington, K. (2003). Alternative stable states in ecology. *Frontiers in Ecology and the Environment*, 1, 376–82.

Bellwood, D. R., Hoey, A. S., and Choat, J. H. (2003). Limited functional redundancy in high diversity systems: resilience and ecosystem function on coral reefs. *Ecology Letters*, 6, 281–5.

Bennett, E. M., Cumming, G. S., and Peterson, G. D. (2005). A systems model approach to determining resilience surrogates for case studies. *Ecosystems*, **8**, 945–57.

Berghöfer, U., Rozzi, R., and Jax, K. (2010). Many eyes on nature: diverse perspectives in the Cape Horn Biosphere Reserve and their relevance for conservation. *Ecology and Society*, **15**(1), 18. [online] URL: http://www.ecologyandsociety.org/vol15/iss1/art18.

Berkes, F., Kislalioglu, M., Folke, C., and Gadgil, M. (1998). Exploring the basic ecological unit: ecosystem-like concepts in traditional societies. *Ecosystems*, **1**, 409–15.

Bertalanffy, L. V. (1968). *General Systems Theory: Foundations, Development, Applications*, New York, NY: Braziller.

Bird, E. A. R. (1987). The social construction of nature: theoretical approaches to the history of environmental problems. *Environmental Review*, **11**, 255–64.

Birnbacher, D. (ed.) (1980). *Ökologie und Ethik*, Stuttgart: Reclam.

Blumrich, H., Bröring, U., Felinks, B., Fromm, H., Mrzljak, J., Schulz, F., Vorwald, J., and Wiegleb, G. (1998). Die Bedeutung der Leitbildentwicklung im Rahmen einer 'guten naturschutzfachlichen Praxis', dargestellt am Beispiel der naturnahen terrestrischen Bereiche der Niederlausitzer Bergbaufolgelandschaften. In *Naturschutz in der Bergbaufolgelandschaft: Leitbildentwicklung*, Landesumweltamt Brandenburg, Potsdam, pp. 1–44.

Boero, F. and Bonsdorff, E. (2007). A conceptual framework for marine biodiversity and ecosystem functioning. *Marine Ecology: An Evolutionary Perspective*, **28**, 134–45.

Borrero, L. A. (1997). The origins of ethnographic subsistence patterns in Fuego-Patagonia. In *Patagonia: Natural History, Prehistory and Ethnography at the Uttermost End of the Earth*, ed. McEwan, C., Borrero, L. A., and Prieto, A. Princeton, NJ: Princeton University Press, pp. 60–81.

Bouamrane, M. (ed.) (2007). *Dialogue in Biosphere Reserves: References, Practices and Experiences*, Paris: UNESCO.

Boughton, D. A., Smith, E. R., and O'Neill, R. V. (1999). Regional vulnerability: a conceptual framework. *Ecosystem Health*, **5**, 312–22.

Boulton, A. J. (1999). An overview of river health assessment: philosophies, practice, problems and prognosis. *Freshwater Biology*, **41**, 469–79.

Boyce, M. S. and Haney, A. (eds) (1997). *Ecosystem Management: Applications for Sustainable Forest and Wildlife Resources*, New Haven, CT/London: Yale University Press.

Boyd, J. and Banzhaf, S. (2007). What are ecosystem services? The need for standardized environmental accounting units. *Ecological Economics*, **63**, 616–26.

Bradshaw, A. D. (1987). Restoration: an acid test for ecology. In *Restoration Ecology: A Synthetic Approach to Ecological Research*, ed. Jordan, W. R., Gilpin, M. E., and Aber, J. D. Cambridge: Cambridge University Press, pp. 23–9.

Bradshaw, A. D. (2002). Introduction and philosophy. In *Handbook of Ecological Restoration. Vol. 1*, ed. Perrow, M. R. and Davy, A. J. Cambridge: Cambridge University Press, pp. 3–9.

Bradshaw, A. D. and Hüttl, R. F. (2001). Future minesite restoration involves a broader approach. *Ecological Engineering*, **17**, 87–90.

Brand, F. (2009a). Resilience and sustainable development: an ecological inquiry. Ph.D. thesis, Technische Universität München.

Brand, F. (2009b). Critical natural capital revisited: ecological resilience and sustainable development. *Ecological Economics*, **68**, 605–12.

Brand, F. S. and Jax, K. (2007). Focussing the meaning(s) of resilience: resilience as a descriptive concept and a boundary object. *Ecology and Society*, **12**(1), 23. [online] URL: http://www.ecologyandsociety.org/vol12/iss1/art23.

Bremner, J. (2008). Species' traits and ecological functioning in marine conservation and management. *Journal of Experimental Marine Biology and Ecology*, **366**, 37–47.

Breymeyer, A. I. (1981). Monitoring of the functioning of ecosystems. *Environmental Monitoring and Assessment*, **1**, 175–83.

Brown, S. and Lugo, A. E. (1994). Rehabilitation of tropical lands: a key to sustaining development. *Restoration Ecology*, **2**, 97–111.

Burke, S. M. and Mitchell, N. (2007). People as ecological participants in ecological restoration. *Restoration Ecology*, **15**, 348–50.

Butler, K. F. and Koontz, T. M. (2005). Theory into practice: implementing ecosystem management objectives in the USDA Forest Service. *Environmental Management*, **35**, 138–50.

Byers, J. E., Cuddington, K., Jones, C. G., Talley, T. S., Hastings, A., Lambrinos, J. G., Crooks, J. A., and Wilson, W. G. (2006). Using ecosystem engineers to restore ecological systems. *Trends in Ecology & Evolution*, **21**, 493–500.

Cairns, J. Jr. (1977). Quantification of biological integrity. In *The Integrity of Waters: Proceedings of a Symposium, March 10–12, 1975*, ed. Ballentine, R. K. and Guarraia, L. J. Washington, DC: US Environmental Agency, pp. 171–87.

Callicott, J. B. (1996). Do deconstructive ecology and sociobiology undermine Leopold's Land Ethic? *Environmental Ethics*, **18**, 353–72.

Callicott, J. B. and Nelson, M. P. (eds) (1998). *The Great New Wilderness Debate*, Athens, GA: University of Georgia Press.

Callicott, J. B., Crowder, L. B., and Mumford, K. (1999). Current normative concepts in conservation. *Conservation Biology*, **13**, 22–35.

Calow, P. (1992). Can ecosystems be healthy? Critical consideration of concepts. *Journal of Aquatic Ecosystem Health*, **1**, 1–5.

Cardinale, B. J., Nelson, K., and Palmer, M. A. (2000). Linking species diversity to the functioning of ecosystems: on the importance of environmental context. *Oikos*, **91**, 175–83.

Cardoso, A. C. (2008). Incorporating invasive alien species into ecological assessment in the context of the Water Framework Directive. *Aquatic Invasions*, **3**, 361–6.

Carlsson, G. (1962). Reflections of functionalism. *Acta Sociologica*, **5**, 201–24.

Carlsson, N. O. L., Brönmark, C., and Hansson, L.-A. (2004). Invading herbivory: the golden apple snail alters ecosystem functioning in Asian wetlands. *Ecology*, **85**, 1575–80.

Carpenter, S., Walker, B., Anderies, J. M., and Abel, N. (2001). From metaphor to measurement: resilience of what to what? *Ecosystems*, **4**, 765–81.

Carpenter, S. R. and Cottingham, K. L. (1997). Resilience and restoration of lakes. *Conservation Ecology*, **1**. [online] http://www.ecologyandsociety.org/vol1/iss1/art2.

Carpenter, S. R. and Lathrop, R. C. (2008). Probabilistic estimate of a threshold for eutrophication. *Ecosystems*, **11**, 601–13.

Carpenter, S. R., Westley, F., and Turner, M. G. (2005). Surrogates for resilience of social–ecological systems. *Ecosystems*, **8**, 941–4.

Carpenter, S. R., Mooney, H. A., Agard, J., Capistrano, D., Defries, R. S., Díaz, S., Dietz, T., Duraiappah, A. K., Oteng-Yeboah, A., Pereira, H. M., Perrings, C., Reid, W. V., Sarukhan, J., Scholes, R. J., and Whyte, A. (2009). Science for managing ecosystem services: beyond the Millennium Ecosystem Assessment. *Proceedings of the National Academy of Sciences of the United States of America*, **106**, 1305–12.

Catovsky, S. (1998). Functional groups: clarifying our use of the term. *Bulletin of the Ecological Society of America*, **79**, 126–7.

Chapin, F. S., Torn, M. S., and Tateno, M. (1996). Principles of ecosystem sustainability. *American Naturalist*, **148**, 1016–37.

Chapman, R. L. (2006). Ecological restoration restored. *Environmental Values*, **15**, 463–78.

Chase, A. (1987). *Playing God in Yellowstone: The Destruction of America's First National Park*, Orlando, FL: Harcourt Brace & Co.

Choi, C. (2008). Tierra del Fuego: the beavers must die. *Nature*, **453**, 968.

Choi, J. S., Frank, K. T., Leggett, W. C., and Drinkwater, K. (2004). Transition to an alternate state in a continental shelf ecosystem. *Canadian Journal of Fisheries and Aquatic Sciences*, **61**, 505–10.

Choi, Y. D. (2007). Restoration ecology to the future: a call for new paradigm. *Restoration Ecology*, **15**, 351–3.

Chovanec, A., Jäger, P., Jungwirth, M., Koller-Kreimel, V., Moog, O., Muhar, S., and Schmutz, S. (2000). The Austrian way of assessing the ecological integrity of running waters: a contribution to the EU Water Framework Directive. *Hydrobiologia*, **422**, 445–52.

Christensen, N. L., Bartuska, A. M., Carpenter, S. R., D'Antonio, C., Francis, R., Franklin, J. F., MacMahon, J. A., Noss, R. F., Parsons, D. J., Peterson, C. H., Turner, M. G., and Woodmansee, R. G. (1996). The report of the Ecological Society of America Committee on the Scientific Basis for Ecosystem Management. *Ecological Applications*, **6**, 665–91.

Clark, T. W. and Zaunbrecher, D. (1987). The Greater Yellowstone Ecosystem: The ecosystem concept in natural resource policy and management. *Renewable Resources Journal*, Summer, 8–16.

Clements, F. E. (1916). *Plant Succession: An Analysis of the Development of Vegetation*, Washington, DC: Carnegie Institution of Washington, Publication No. 242.

Clewell, A. F. and Aronson, J. (2006). Motivations for the restoration of ecosystems. *Conservation Biology*, **20**, 420–8.

Clewell, A. and McDonald, T. (2009). Relevance of natural recovery to ecological restoration. *Ecological Restoration*, **27**, 122–4.

Colautti, R. J. and MacIsaac, H. J. (2004). A neutral terminology to define 'invasive' species. *Diversity and Distributions*, **10**, 135–41.

Cole, D. N. (2000). Paradox of the primeval: ecological restoration in wilderness. *Ecological Restoration*, **18**, 77–86.

Cole, D. N. and Landres, P. B. (1996). Threats to wilderness ecosystems: impacts and research needs. *Ecological Applications*, **6**, 168–84.

Cole, G. F. (1968). *Elk and the Primary Purpose of Yellowstone National Park*, Manuscript, Yellowstone National Park Library.

Cole, G. F. (1969). *Elk and the Yellowstone Ecosystem*, Manuscript, Yellowstone National Park Library.

Connell, J. H. and Sousa, W. P. (1983). On the evidence needed to judge ecological stability or persistence. *American Naturalist*, **121**, 789–824.

Convention on Biological Diversity (2004). *The Ecosystem Approach (CBD Guidelines)*, Montreal, QC: Secretariat of the Convention on Biological Diversity.

Cooke, B. and Kothari, U. (eds) (2001). *Participation: The New Tyranny?*, London: Zed Books.

Cooper, D. J. (2008). Is Yellowstone on the brink? *Ecology*, **89**, 293–4.

Costanza, R. (1992). Toward an operational definition of ecosystem health. In *Ecosystem Health: New Goals for Environmental Management*, ed. Costanza, R., Norton, B. G., and Haskell, B. D. Washington, DC: Island Press, pp. 239–56.

Costanza, R., D'Arge, R., de Groot, R., Farber, S., Grasso, M., Hannon, B., Limburg, K., Naeem, S., O'Neill, R. V., Paruelo, J., Raskin, R. G., Sutton, P., and Van Den Belt, M. (1997). The value of the world's ecosystem services and natural capital. *Nature*, **387**, 253–60.

Cousins, S. H. (1987). Can we count natural ecosystems? *Bulletin of the British Ecological Society*, **18**, 156–8.

Couve, E. and Vidal-Ojeda, C. (2000). *Birds of the Beagle Channel: Cape Horn, Staten Island, Diego Ramírez Islands and Surrounding Seas*, Punta Arenas: Fantastico Sur.

Crist, E. (2004). Against the social construction of nature and wilderness. *Environmental Ethics*, **26**, 5–24.

Cronon, W. (ed.) (1995). *Uncommon Ground: Rethinking the Human Place in Nature*, New York, NY /London: W. W. Norton & Co.

Cumming, G. S. and Collier, J. (2005). Change and identity in complex systems. *Ecology and Society*, **10**(1), 29. [online] URL: http://www.ecologyandsociety.org/vol10/iss1/art29.

Cumming, G. S., Barnes, G., Perz, S., Schmink, M., Sieving, K. E., Southworth, J., Binford, M., Holt, R. D., Stickler, C., and Van Holt, T. (2005). An exploratory framework for the empirical measurement of resilience. *Ecosystems*, **8**, 975–87.

Cummins, K. W. (1974). Structure and function of stream ecosystems. *BioScience*, **24**, 631–41.

Daily, G. C. (ed.) (1997). *Nature's Services: Societal Dependence on Natural Ecosystems*, Washington, DC: Island Press.

Danovaro, R., Gambi, C., Dell'anno, A., Corinaidesi, C., Fraschetti, S., Vanreusel, A., Vincx, M., and Gooday, A. J. (2008). Exponential decline of deep-sea ecosystem functioning linked to benthic biodiversity loss. *Current Biology*, **18**, 1–8.

de Groot, R. S. (1987). Environmental functions as a unifying concept for ecology and economics. *The Environmentalist*, **7**, 105–9.

de Groot, R. S. (1992). *Functions of Nature: Evaluation of Nature in Environmental Planning, Management and Decision-making*, Groningen: Wolters Noordhoff BV.

de Groot, R. S., Wilson, M. A., and Boumans, R. M. J. (2002). A typology for the classification, description and valuation of ecosystem functions, goods and services. *Ecological Economics*, **41**, 393–408.

de Groot, W. T. and Van Den Born, R. J. G. (2003). Visions of nature and landscape type preferences: an exploration in the Netherlands. *Landscape and Urban Planning*, **63**, 127–38.

de Jonge, V. N. (2007). Toward the application of ecological concepts in EU coastal water management. *Marine Pollution Bulletin*, **55**, 407–14.

de Leo, G. A. and Levin, S. (1997). The multifaceted aspects of ecosystem integrity. *Conservation Ecology*, **1**, 3. [online] http://www.ecologyandsociety. org/vol1/iss1/art3.

Delcourt, H. R. and Delcourt, P. A. (1991). *Quaternary Ecology. A Paleoecological Perspective*, London: Chapman & Hall.

Demeritt, D. (2002). What is the 'social construction of nature'? A typology and sympathetic critique. *Progress in Human Geography*, **26**, 767–90.

Dent, C. L., Cumming, G. S., and Carpenter, S. R. (2002). Multiple states in river and lake ecosystems. *Philosophical Transactions of the Royal Society London B*, **357**, 635–45.

Desjardins, J. R. (2006). *Environmental Ethics: An Introduction to Environmental Philosophy*, Boston, MA: Wadsworth.

Despain, D., Houston, D., Meagher, M., and Schullery, P. (1986). *Wildlife in Transition: Man and Nature on Yellowstone's Northern Range*, Boulder, CO: Roberts Rinehart.

DeYoung, B., Barange, M., Beaugrand, G., Harris, R., Perry, R. I., Scheffer, M., and Werner, F. (2008). Regime shifts in marine ecosystems: detection, prediction and management. *Trends in Ecology & Evolution*, **23**, 402–9.

Díaz, S., Symstad, A. J., Chapin, F. S. I., Wardle, D. A., and Huenneke, L. F. (2003). Functional diversity revealed by removal experiments. *Trends in Ecology & Evolution*, **18**, 140–6.

Didham, R. K., Watts, C. H., and Norton, D. A. (2005). Are systems with strong underlying abiotic regimes more likely to exhibit alternative stable states? *Oikos*, **110**, 409–16.

Dietz, T. and Stern, P. C. (eds) (2008). *Public Participation in Environmental Assessment and Decision Making*, Washington, DC: National Academies Press.

Duffy, J. E., Richardson, J. P., and Canuel, E. A. (2003). Grazer diversity effects on ecosystem functioning in seagrass beds. *Ecology Letters*, **6**, 637–45.

Dufour, S. and Piégay, H. (2009). From the myth of a lost paradise to targeted river restoration: forget natural references and focus on human benefits. *River Research and Applications*, **25**, 568–81.

Dupré, J. (1992). Species: theoretical contexts. In *Keywords in Evolutionary Biology*, ed. Keller, E. F. and Lloyd, E. A. Cambridge, MA: Harvard University Press, pp. 312–17.

Ecological Society of America (1999). *Biodiversity and Ecosystem Functioning: Maintaining Natural Life Support Systems (Issues in Ecology, Vol. 4)*, Washington, DC: Ecological Society of America.

Ecological Society of America (2008). The role of ecosystem services in conservation and resource management. *Frontiers in Ecology and the Environment*, **7**, 1–60.

Ehrenfeld, J. G. (2000). Defining the limits of restoration: the need for realistic goals. *Restoration Ecology*, **8**, 2–9.

Ehrlich, P. and Ehrlich, A. (1981). *Extinction: The Causes and Consequences of the Disappearance of Species*, New York, NY: Random House.

Ehrlich, P. and Walker, B. H. (1998). Rivets and redundancy. *BioScience*, **48**, 387.

Elmqvist, T., Folke, C., Nyström, M., Peterson, G., Bengtsson, J., Walker, B., and Norberg, J. (2003). Response diversity, ecosystem change, and resilience. *Frontiers in Ecology and the Environment*, **1**, 488–94.

Elton, C. S. (1927). *Animal Ecology*, London: Sidgwick & Jackson.

Elton, C. S. (1958). *The Ecology of Invasions by Animals and Plants*, London: Methuen & Co.

Eser, U. (1999). *Der Naturschutz und das Fremde: Ökologische und normative Grundlagen der Umweltethik*, Frankfurt: Campus.

Eser, U. (2002). Der Wert der Vielfalt: "Biodiversität" zwischen Wissenschaft, Politik und Ethik. In *Umwelt – Ethik – Recht*, ed. Bobbert, M., Düwell, M., and Jax, K. Tübingen: Francke-Verlag, pp. 160–81.

Eser, U. and Potthast, T. (1999). *Naturschutzethik: Eine Einführung für die Praxis*, Baden-Baden: Nomos.

Essler, W. K. (1982). *Wissenschaftstheorie I: Definition und Reduktion*, Freiburg/München: Alber.

European Commission (2003). *Rivers and Lakes: Typology, Reference Conditions and Classification Systems. Common Implementation Strategy for the Water Framework Directive (2000/60/EC), Guidance Document No. 10*, Luxembourg: Office for Official Publications of the European Communities. [online] http://circa.europa.eu/Public/irc/env/wfd/library?l=/framework_directive/guidance_documents.

European Commission (2005). *Overall Approach to the Classification of Ecological Status and Ecological Potential. Common Implementation Strategy for the Water Framework Directive (2000/60/EC), Guidance Document No. 13*, Luxembourg: Office for Official Publications of the European Communities. [online] http://circa.europa.eu/Public/irc/env/wfd/library?l=/framework_directive/guidance_documents.

European Community (2000). Directive 2000/60/EC of 23 October 2000 of the European Parliament and of the Council establishing a framework for Community action in the field of water policy. *Official Journal of the European Communities*, **L 327**, 1–72.

Felinks, B. and Wiegleb, G. (1998). Welche Dynamik schützt der Prozeßschutz? Aspekte unterschiedlicher Maßstabsebenen: dargestellt am Beispiel der Niederlausitzer Bergbaufolgelandschaft. *Naturschute und Landschaftsplanung*, **30**, 298–303.

Fischer, A. and Young, J. C. (2007). Understanding mental constructs of biodiversity: implications for biodiversity management and conservation. *Biological Conservation*, **136**, 271–82.

Fisher, B., Turner, R. K., and Morling, P. (2009). Defining and classifying ecosystem services for decision making. *Ecological Economics*, **68**, 643–53.

Flahault, C. and Schröter, C. (eds) (1910). *Phytogeographische Nomenklatur: III. Internationaler Botanischer Kongress, Brüssel 1910*, Zürich: Zürcher & Furrer.

244 · References

Folke, C. (2006). Resilience: the emergence of a perspective for social–ecological systems analyses. *Global Environmental Change*, **16**, 253–67.

Folke, C., Carpenter, S. R., Elmqvist, T., Gunderson, L. H., Holling, C. S., and Walker, B. (2002). Resilience and sustainable development: building adaptive capacity in a world of transformations. *Ambio*, **31**, 437–40.

Forman, R. T. T. and Godron, M. (1986). *Landscape Ecology*, New York, NY: John Wiley & Sons.

Fox, D. (2007). Back to the no-analog future? *Science*, **316**, 823–4.

Free, A. and Barton, N. H. (2007). Do evolution and ecology need the Gaia hypothesis? *Trends in Ecology & Evolution*, **22**, 611–19.

Gamfeldt, L. and Hillebrand, H. (2008). Biodiversity effects on aquatic ecosystem functioning: maturation of a new paradigm. *International Review of Hydrobiology*, **93**, 550–64.

Gamfeldt, L., Hillebrand, H., and Jonsson, P. R. (2008). Multiple functions increase the importance of biodiversity for overall ecosystem functioning. *Ecology*, **89**, 1223–31.

Gandy, M. (1996). Crumbling land: the postmodernity debate and the analysis of environmental problems. *Progress in Human Geography*, **20**, 23–40.

Gaston, K. J. (1996). What is biodiversity? In *Biodiversity: A Biology of Numbers and Difference*, ed. Gaston, K. J. Oxford: Blackwell, pp. 1–9.

Gattie, D. K., Smith, M. C., Tollner, E. W., and McCutcheon, S. C. (2003). The emergence of ecological engineering as a discipline. *Ecological Engineering*, **20**, 409–20.

Giller, P. S., Hillebrand, H., Berninger, U.-G., Gessner, M. O., Hawkins, S., Inchausti, P., Inglis, C., Leslie, H., Malmqvist, B., Monaghan, M. T., Morin, P. J., and O'Mullan, G. (2004). Biodiversity effects on ecosystem functioning: emerging issues and their experimental test in aquatic environments. *Oikos*, **104**, 423–36.

Gindele, M. (1999). Die Funktion der Biodiversität: zur Problematik der Redundanz von Arten in Ökologie und Naturschutz. *Landschaftsentwicklung und Umweltforschung: Schriftenreihe im Fachbereich Umwelt und Gesellschaft der TU Berlin*, 112.

Gitay, H., Wilson, J. B., and Lee, W. G. (1996). Species redundancy: a redundant concept? *Journal of Ecology*, **84**, 121–4.

Glacken, C. J. (1967). *Traces on the Rhodian Shore: Nature and Culture in Western Thought from Ancient Times to the End of the Eighteenth Century*, Berkeley, CA: University of California Press.

Gleason, H. A. (1917). The structure and development of the plant association. *Bulletin of the Torrey Botanical Club*, **44**, 463–81.

Gleason, H. A. (1926). The individualistic concept of the plant association. *Bulletin of the Torrey Botanical Club*, **53**, 7–26.

Glick, D., Carr, M., and Harting, B. (1991). *An Environmental Profile of the Greater Yellowstone Ecosystem*, Bozeman, MT: Greater Yellowstone Coalition.

Gobster, P. H. and Hull, R. B. (2000). *Restoring Nature: Perspectives from the Social Sciences and Humanities*, Washington, DC: Island Press.

Golley, F. B. (1993). *A History of the Ecosystem Concept in Ecology: More than the Sum of its Parts*, New Haven, CT: Yale University Press.

Goodman, D. (1975). The theory of diversity–stability relationships in ecology. *The Quarterly Review of Biology*, **50**, 237–66.

Gorke, M. (2007). Eigenwert der Natur: Ethische Begründung und Konsequenzen. Habilitationsschrift. Mathematisch-Naturwissenschaftliche Fakultät der Universität Greifswald.

Grant, W. E. and French, N. R. (1980). Evaluation of the role of small mammals in grassland ecosystems: a modelling approach. *Ecological Modelling*, **8**, 15–37.

Greater Yellowstone Coordinating Committee (1990). *Vision for the Future: A Framework for Coordination in the Greater Yellowstone Area*. Draft, Billings, MT: Greater Yellowstone Coordinating Committee.

Griffin, J. N., O'Gorman, E. J., Emmerson, M. C., Jenkins, S. R., Klein, A.-M., Loreau, M., and Symstad, A. J. (2009). Biodiversity and the stability of ecosystem functioning. In *Biodiversity, Ecosystem Functioning, & Human Wellbeing: An Ecological and Economic Perspective*, ed. Naeem, S., Bunker, D. E., Hector, A., Loreau, M., and Perrings, C. Oxford: Oxford University Press, pp. 78–93.

Griffiths, B. S., Ritz, K., Bardgett, R. D., Cook, R., Christensen, S., Ekelund, F., Sorensen, S. J., Baath, E., Bloem, J., de Ruiter, P. C., Dolfing, J., and Nicolardot, B. (2000). Ecosystem response of pasture soil communities to fumigation-induced microbial diversity reductions: an examination of the biodiversity-ecosystem function relationship. *Oikos*, **90**, 279–94.

Grime, J. P. (1979). *Plant Strategies and Vegetation Processes*, Chichester: Wiley & Sons.

Grimm, V. and Wissel, C. (1997). Babel, or the ecological stability discussions: an inventory and analysis of terminology and a guide for avoiding confusion. *Oecologia*, **109**, 323–34.

Grisebach, A. (1838). Über den Einfluß des Klimas auf die Begrenzung der natürlichen Floren. In *Gesammelte Abhandlungen und kleinere Schriften zur Pflanzengeographie*, ed. Grisebach, A. Leipzig: Verlag von Wilhelm Engelmann, pp. 1–29.

Gross, M. (2003). *Inventing Nature: Ecological Restoration by Public Experiments*, Lanham, MD: Lexington Books.

Großer, K.-H. (1995). Das Lausitzer Braunkohlerevier: Der Naturraum und seine Gestaltung. In *Braunkohletagebau und Rekultivierung*, ed. Pflug, W. Berlin/Heidelberg/New York, NY: Springer, pp. 461–74.

Grumbine, R. E. (1994a). Conservation biology in context: an interview with Michael Soulé. In *Environmental Policy and Biodiversity*, ed. Grumbine, R. E. Washington, DC: Island Press, pp. 99–105.

Grumbine, R. E. (1994b). What is ecosystem management? *Conservation Biology*, **8**, 27–38.

Gunderson, L. H. (2000). Ecological resilience: in theory and application. *Annual Review of Ecology, Evolution, and Systematics*, **31**, 425–39.

Gunderson, L. H. and Holling, C. S. (eds) (2002). *Panarchy: Understanding Transformations in Human and Natural Systems*, Washington, DC: Island Press.

Gusinde, M. (1937). *Die Feuerland-Indianer. Vol. 2: Die Yamana: vom Leben und Denken der Wassernomaden am Kap Horn*, Mödling bei Wien: Verlag der Internationalen Zeitschrift 'Anthropos'.

Haber, W. (1979). Theoretische Anmerkungen zu 'ökologischen Planung'. *Verhandlungen der Gesellschaft für Ökologie*, **7**, 19–30.

Hacking, I. (1999). *The Social Construction of What?*, Cambridge, MA/London: Harvard University Press.

Häge, K. (1996). Recultivation in the Lusatian mining region: targets and prospects. *Water Air and Soil Pollution*, **91**, 43–57.

Haider, S. and Jax, K. (2007). The application of environmental ethics in biological conservation: a case study from the southernmost tip of the Americas. *Biodiversity and Conservation*, **16**, 2559–73.

Hampicke, U. (1993). Naturschutz und Ethik: Rückblick auf eine 20jährige Diskussion, 1973–1993, und politische Folgerungen. *Zeitschrift für Ökologie und Naturschutz*, **2**, 73–86.

Hard, G. (1969). 'Kosmos' und 'Landschaft': Kosmologische und landschafts-physiognomische Denkmotive bei Alexander von Humboldt und in der geographischen Humboldt-Auslegung des 20. Jahrhunderts. In *Alexander von Humboldt. Werk und Weltgeltung*, ed. Pfeiffer, H. München: Piper-Verlag, pp. 133–77.

Hargrove, E. C. (1989). *Foundations of Environmental Ethics*, Englewood Cliffs, NJ: Prentice Hall.

Harper, J. L. (1977). *Population Biology of Plants*, London: Academic Press.

Harper, J. L. and Hawksworth, D. L. (1994). Biodiversity: measurement and estimation. *Philosophical Transactions of the Royal Society B*, **345**, 5–12.

Harris, J. A., Hobbs, R. J., Higgs, E. S., and Aronson, J. (2006). Ecological restoration and climate change. *Restoration Ecology*, **14**, 170–6.

Hartje, V., Klaphake, A., and Schliep, R. (2003). *The International Debate on the Ecosystem Approach: Critical Review, International Actors, Obstacles and Challenges*, Bonn: Bundesamt für Naturschutz.

Haselhuhn, I. and Leßmann, D. (2005). *Rahmenbedingungen der Entwicklung der Lausitzer Bergbaulandschaft, insbesondere der Entstehung und des Managements von Tagebauseen. Gewässerreport Nr. 9. BTU Cottbus, Aktuelle Reihe 2/2005*, Cottbus: TU Cottbus.

Haskell, B. D., Norton, B. G., and Costanza, R. (1992). What is ecosystem health and why should we worry about it? In *Ecosystem Health: New Goals for Environmental Management*, ed. Costanza, R., Norton, B. G., and Haskell, B. D. Washington, DC: Island Press, pp. 3–20.

Hättenschwiler, S., Tiunov, A. V., and Scheu, S. (2005). Biodiversity and litter decomposition in terrestrial ecosystems. *Annual Review of Ecology, Evolution and Systematics*, **36**, 191–218.

Hatton-Ellis, T. (2008). The hitchhiker's guide to the Water Framework Directive. *Aquatic Conservation: Marine and Freshwater Ecosystems*, **18**, 111–16.

Haubold-Rosar, M. (2002). Landwirtschaft. In *Wissenschaftliche Begleitung der ostdeutschen Braunkohlensanierung: Forschungsprojekte 1994–2000*, ed. Lmbv. Berlin: LMBV Eigenverlag, pp. 132–54.

Haubold-Rosar, M. (2004). Rekultivierung von Braunkohlenbergbauflächen für eine nachhaltige landwirtschaftliche Nutzung: Bisherige Forschungsergebnisse und aktueller Forschungsbedarf. *Zeitschrift für Angewandte Umweltforschung (ZAU)*, Sonderheft **14**, 158–68.

Hector, A. and Bagchi, R. (2007). Biodiversity and ecosystem multifunctionality. *Nature*, **448**, 188–91.

Hector, A., Bell, T., Connolly, J., Finn, J., Fox, J., Kirwan, L., Loreau, M., McLaren, J., Schmid, B., and Weigelt, A. (2009). The analysis of biodiversity experiments: from pattern toward mechanism. In *Biodiversity, Ecosystem Functioning, & Human Wellbeing: An Ecological and Economic Perspective*, ed. Naeem, S., Bunker, D. E., Hector, A., Loreau, M., and Perrings, C. Oxford: Oxford University Press, pp. 94–104.

Heger, T. and Trepl, L. (2003). Predicting biological invasions. *Biological Invasions*, **5**, 313–21.

Heger, T. and Trepl, L. (2008). Was sind invasive gebietsfremde Arten? *Natur und Landschaft*, **83**, 399–401.

Heiskanen, A. S., Van de Bund, W., Cardoso, A. C., and Nõges, P. (2004). Towards good ecological status of surface waters in Europe: interpretation and harmonisation of the concept. *Water Science and Technology*, **49**, 169–77.

Hempel, C. G. (1952). *Fundamentals of Concept Formation in Empirical Science*, Chicago, IL: University of Chicago Press.

Hering, D., Feld, C. K., Moog, O., and Ofenbock, T. (2006). Cook book for the development of a Multimetric Index for biological condition of aquatic ecosystems: experiences from the European AQEM and STAR projects and related initiatives. *Hydrobiologia*, **566**, 311–24.

Hesse, H. (2000). Vom Zweck zur Funktion: Hinweise aus wissenschaftsphilosophischer Sicht. In *Funktionsbegriff und Unsicherheit in der Ökologie*, ed. Jax, K. Frankfurt: Peter Lang, pp. 19–30.

Higgs, E. (1997). What is good ecological restoration? *Conservation Biology*, **11**, 338–48.

Higgs, E. (2003). *Nature by Design: People, Natural Processes and Ecological Restoration*, Cambridge, MA: MIT Press.

Hilderbrand, R. H., Watts, A. C., and Randle, A. M. (2005). The myths of restoration ecology. *Ecology and Society*, **10**(1), 19.

Hobbs, R. J. and Norton, D. A. (1996). Towards a conceptual framework for restoration ecology. *Restoration Ecology*, **4**, 93–110.

Hobbs, R. J., Davis, M. A., Slobodkin, L. B., Lackey, R. T., Halvorson, W., and Throop, W. (2004). Restoration ecology: the challenge of social values and expectations (Forum). *Frontiers in Ecology and the Environment*, **2**, 43–8.

Hobbs, R. J., Arico, S., Aronson, J., Baron, J. S., Bridgewater, P., Cramer, V. A., Epstein, P. R., Ewel, J. J., Klink, C. A., Lugo, A. E., Norton, D., Ojima, D., Richardson, D. M., Sanderson, E. W., Valladares, F., Vila, M., Zamora, R., and Zobel, M. (2006). Novel ecosystems: theoretical and management aspects of the new ecological world order. *Global Ecology and Biogeography*, **15**, 1–7.

Hobbs, R. J., Higgs, E., and Harris, J. A. (2009). Novel ecosystems: implications for conservation and restoration. *Trends in Ecology & Evolution*, **24**, 599–605.

Holling, C. S. (1973). Resilience and stability of ecological systems. *Annual Review of Ecology, Evolution, and Systematics*, **4**, 1–23.

Holling, C. S. (1996). Engineering resilience versus ecological resilience. In *Engineering with Social Constraints*, ed. Schulze, P. C. Washington, DC: National Academy Press, pp. 31–43.

Hooper, D. U., Solan, M., Symstad, A. J., Díaz, S., Gessner, O., Buchmann, N., Degrange, V., Grime, P., Hulot, F., Mermillod-Blondin, F., Roy, J., Spehn, E., and Van Peer, L. (2002). Species diversity, functional diversity, and ecosystem functioning. In *Biodiversity and Ecosystem Functioning: Synthesis and Perspectives*, ed. Loreau, M., Naeem, S., and Inchausti, P. Oxford: Oxford University Press, pp. 195–208.

Hooper, D. U., Chapin, F. S. I., Ewel, J. J., Hector, A., Inchausti, P., Lavorel, S., Lawton, J. H., Lodge, D. M., Loreau, M., Naeem, S., Schmid, B., Setala, H., Symstad, A. J., Vandermeer, J., and Wardle, D. A. (2005). Effects of biodiversity on ecosystem functioning: a consensus of current knowledge. *Ecological Monographs*, **75**, 3–35.

Hoppe, B. (1990). Physiognomik der Vegetation zur Zeit von Alexander von Humboldt. In *Alexander von Humboldt. Weltbild und Wirkung auf die Wissenschaften*, ed. Lindgren, U. Köln, Wien: Böhlau, pp. 77–102.

Hornborg, A. (2009). Zero-sum world: challenges in conceptualizing environmental load displacement and ecologically unequal exchange in the world-system. *International Journal of Comparative Sociology*, **50**, 237–62.

Houston, D. B. (1971). Ecosystems of national parks. *Science*, **172**, 648–51.

Houston, D. B. (1981). Yellowstone elk: some thoughts on experimental management. *Pacific Park Science*, **1**, 4–6.

Houston, D. B. (1982). *The Northern Yellowstone Elk: Ecology and Management*, New York, NY: Macmillan.

Huff, D. E. and Varley, J. D. (1999). Natural regulation in Yellowstone National Park's northern range. *Ecological Applications*, **9**, 17–29.

Humboldt, A. von (1807 [1969]). *Ansichten der Natur*, Stuttgart: Reclam.

Hutchinson, G. E. (1957). Concluding remarks. *Cold Spring Harbor Symposia on Quantitative Biology*, **22**, 415–27.

Hutchinson, G. E. (1978). *An Introduction to Population Ecology*, New Haven, CT/ London: Yale University Press.

Hüttl, R. F. (1998). Ecology of post strip-mining landscapes in Lusatia, Germany. *Environmental Science & Policy*, **1**, 129–35.

Hüttl, R. F. (2000). Forstliche Rekultivierung im Lausitzer Braunkohlerevier. *Rundgespräche der Kommission für Ökologie*, **20**, 53–64.

Hüttl, R. F. and Weber, E. (2001). Forest ecosystem development in post-mining landscapes: a case study of the Lusatian lignite district. *Naturwissenschaften*, **88**, 322–9.

Ibelings, B. W., Portielje, R., Lammens, E. H. R. R., Noordhuis, R., Van Den Berg, M. S., Joosse, W., and Meijer, M. L. (2007). Resilience of alternative stable states during the recovery of shallow lakes from eutrophication: Lake Veluwe as a case study. *Ecosystems*, **10**, 4–16.

IPCC (2001). *Climate Change 2001: Synthesis Report*, Cambridge: Cambridge University Press.

IPCC (2008). *Climate Change 2007: Synthesis Report*, Geneva: IPCC.

Ives, A. R. and Carpenter, S. R. (2007). Stability and diversity of ecosystems. *Science*, **317**, 58–62.

Jaksic, F. M., Iriate, J. A., Jiménez, J. E., and Martínez, D. R. (2002). Invaders without frontiers: cross-border invasions of exotic mammals. *Biological Invasions*, **4**, 157–73.

Janzen, D. H. (1977). What are dandelions and aphids? *American Naturalist*, **111**, 586–9.

Jasanoff, S. (2004). Ordering knowledge, ordering society. In *States of Knowledge: The Co-production of Science and Social Order*, ed. Jasanoff, S. London: Routledge, pp. 13–45.

Jax, K. (1998). Holocoen and ecosystem: on the origin and historical consequences of two concepts. *Journal of the History of Biology*, **31**, 113–42.

Jax, K. (2000). Verschiedene Verständnisse des Funktionsbegriffs in den Umweltwissenschaften. In *Funktionsbegriff und Unsicherheit in der Ökologie*, ed. Jax, K. Frankfurt/Berlin: Peter Lang, pp. 7–17.

Jax, K. (2001). Naturbild, Ökologietheorie und Naturschutz: zur Geschichte des Ökosystemmanagements im Yellowstone-Nationalpark. *Verhandlungen zur Geschichte und Theorie der Biologie*, **7**, 115–34.

Jax, K. (2002a). *Die Einheiten der Ökologie. Analyse, Methodenentwicklung und Anwendung in Ökologie und Naturschutz*, Frankfurt: Peter Lang.

Jax, K. (2002b). Zur Transformation ökologischer Fachbegriffe beim Eingang in Verwaltungsnormen und Rechtstexte: das Beispiel des Ökosystem-Begriffs. In *Umwelt, Ethik & Recht*, ed. Bobbert, M., Düwell, M., and Jax, K. Tübingen: Francke-Verlag, pp. 69–97.

Jax, K. (2003). Wofür braucht der Naturschutz die wissenschaftliche Ökologie? Die Kontroversen um den Hudson River als Testfall. *Natur und Landschaft*, **78**, 93–9.

Jax, K. (2005). Function and 'functioning' in ecology? What does it mean? *Oikos*, **111**, 641–8.

Jax, K. (2006). The units of ecology: definitions and application. *Quarterly Review of Biology*, **81**, 237–58.

Jax, K. (2007). Can we define ecosystems? On the confusion between definition and description of ecological concepts. *Acta Biotheoretica*, **55**, 341–55.

Jax, K. and Rozzi, R. (2004). Ecological theory and values in the determination of conservation goals: examples from the temperate regions of Germany, USA and Chile. *Revista Chilena de Historia Natural*, **77**, 349–66.

Jax, K. and Schwarz, A. (2010). Structure of the handbook. In *Revisiting Ecology: Reviewing Concepts, Advancing Science*, ed. Schwarz, A. and Jax, K. Dordrecht: Springer.

Jax, K., Jones, C. G., and Pickett, S. T. A. (1998). The self-identity of ecological units. *Oikos*, **82**, 253–64.

Jeltsch, F., Weber, G. E., and Grimm, V. (2000). Ecological buffering mechanisms in savannas: a unifying theory of long-term tree–grass coexistence. *Plant Ecology*, **150**, 161–71.

Jetzkowitz, J. and Stark, C. (2003). Einführung: Der Funktionalismus und die Frage nach der Methodologie. In *Soziologischer Funktionalismus. Zur Methodologie einer Theorietradition*, ed. Jetzkowitz, J., and Stark, C. Opladen: Leske & Budrich, pp. 7–16.

Jones, C. G., Lawton, J. H., and Shachak, M. (1994). Organisms as ecosystem engineers. *Oikos*, **69**, 373–86.

Jones, C. G., Lawton, J. H., and Shachak, M. (1997). Positive and negative effects of organisms as physical ecosystem engineers. *Ecology*, **78**, 1946–57.

Jordan, W. R., Gilpin, M. E., and Aber, J. D. (1987). Restoration ecology: ecological restoration as a technique to basic research. In *Restoration Ecology: A Synthetic Approach to Ecological Research*, ed. Jordan, W. R., Gilpin, M. E., and Aber, J. D. Cambridge: Cambridge University Press, pp. 3–21.

Jørgensen, S. E. (1992). Exergy and ecology. *Ecological Modelling*, **63**, 185–214.

Jørgensen, S. E., Patten, B. C., and Straskraba, M. (1992). Ecosystems emerging: toward an ecology of complex systems in a complex future. *Ecological Modelling*, **62**, 1–27.

Junker, B., Buchecker, M., and Müller-Böker, U. (2007). Objectives of public participation: which actors should be involved in the decision making for river restorations? *Water Resources Research*, **43**(10), W10438.

Kant, I. (1790 [2009]). *Kritik der Urteilskraft*, Frankfurt am Main: Suhrkamp.

Karr, J. R. (1991). Biological integrity: a long-neglected aspect of water resource management. *Ecological Applications*, **1**, 66–84.

Karr, J. R. (1993). Defining and assessing ecological integrity: beyond water quality. *Environmental Toxicology and Chemistry*, **12**, 1521–31.

Karr, J. R. (1996). Ecological integrity and ecological health are not the same. In *Engineering with Social Constraints*, ed. Schulze, P. C. Washington, DC: National Academy Press, pp. 97–109.

Karr, J. R. (2000). Health, integrity, and biological assessment: the importance of measuring whole things. In *Ecological Integrity: Integrating Environment, Conservation, and Health*, ed. Pimentel, D., Westra, L., and Noss, R. F. Washington, DC: Island Press, pp. 209–26.

Karr, J. R. and Dudley, D. R. (1981). Ecological perspective in water quality goals. *Environmental Management*, **5**, 55–68.

Kay, C. E. (1995). Aboriginal overkill and native burning: implications for modern ecosystem management. In *Sustainable Society and Protected Areas: Contributed Papers of the 8th Conference on Research and Resource Management in Parks and on Public Lands, April 17–21, 1995, Portland, Oregon*, ed. Linn, R. Hancock, MI: The George Wright Society, pp. 107–18.

Kay, C. E. (1997). Viewpoint: ungulate herbivory, willows, and political ecology in Yellowstone. *Journal of Range Management*, **50**, 139–45.

Kay, J. J. (1991). A nonequilibrium thermodynamic framework for discussing ecosystem integrity. *Environmental Management*, **15**, 483–95.

Kay, J. J. (1993). On the nature of ecological integrity: some closing comments. In *Ecological Integrity and the Management of Ecosystems*, ed. Woodley, S., Kay, J., and Francis, G. Delray Beach, FL: St. Lucie Press, pp. 201–12.

Keiter, R. B. and Boyce, M. S. (eds) (1991). *The Greater Yellowstone Ecosystem: Redefining America's Wilderness Heritage*, New Haven, CT: Yale University Press.

Keller, E. F. (2005). Ecosystems, organisms, and machines. *BioScience*, **55**, 1069–74.

Kidner, D. W. (2000). Fabricating nature: a critique of the social construction of nature. *Environmental Ethics*, **22**, 339–57.

King, A. W. (1993). Considerations of scale and hierarchy. In *Ecological Integrity and the Management of Ecosystems*, ed. Woodley, S., Kay, J., and Francis, G. Delray Beach, FL: St. Lucie Press, pp. 19–45.

King, E. G. and Hobbs, R. J. (2006). Identifying linkages among conceptual models of ecosystem degradation and restoration: towards an integrative framework. *Restoration Ecology*, **14**, 369–78.

Kinzig, A. P., Pacala, S. W., and Tilman, D. (eds) (2001). *The Functional Consequences of Biodiversity. Empirical Progress and Theoretical Extensions*, Princeton, NJ/Oxford: Princeton University Press.

Kirchhoff, T., Brand, F., Hoheisel, D., and Grimm, V. (2010). The one-sidedness and cultural bias of the resilience approach. *Gaia*, **19**, 25–32.

Kitcher, P. (2001). *Science, truth, and democracy*, Oxford/New York, NY: Oxford University Press.

Klijn, F. and Udo de Haes, H. A. (1994). A hierarchical approach to ecosystems and its implication for ecological land classification. *Landscape Ecology*, **9**, 89–104.

Knops, J. M. H., Tilman, D., Haddad, N. M., Naeem, S., Mitchell, C. E., Haarstad, J., Ritchie, M. E., Howe, K. M., Reich, P. B., Siemann, E., and Groth, J. (1999). Effects of plant species richness on invasion dynamics, disease outbreaks, insect abundances and diversity. *Ecology Letters*, **2**, 286–93.

Knowlton, N. (1992). Thresholds and multiple stable states in coral-reef community dynamics. *American Zoologist*, **32**, 674–82.

Kremen, C. (2005). Managing ecosystem services: what do we need to know about their ecology? *Ecology Letters*, **8**, 468–79.

Kremen, C. and Ostfeld, R. S. (2005). A call to ecologists: measuring, analyzing, and managing ecosystem services. *Frontiers in Ecology and the Environment*, **3**, 540–8.

Lackey, R. T. (1998). Seven pillars of ecosystem management. *Landscape and Urban Planning*, **40**, 21–30.

Lackey, R. T. (2001). Values, policy, and ecosystem health. *Bioscience*, **51**, 437–43.

Larsen, T. H., Williams, N M., and Kremen, C. (2005). Extinction order and altered community structure rapidly disrupt ecosystem functioning. *Ecology Letters*, **8**, 538–47.

Larson, B. M. H. (2005). The war of the roses: demilitarizing invasion biology. *Frontiers in Ecology and the Environment*, **3**, 495–500.

Latour, B. (1999). *Pandora's Hope: Essays on the Reality of Science Studies*, Cambridge, MA: Harvard University Press.

Lavorel, S. and Garnier, E. (2002). Predicting changes in community composition and ecosystem functioning from plant traits: revisiting the Holy Grail. *Functional Ecology*, **16**, 545–56.

Lawler, S. P., Armesto, J. J., and Kareiva, P. (2001). How relevant to conservation are studies linking biodiversity and ecosystem functioning? In *The Functional Consequences of Biodiversity: Empirical Progress and Theoretical Extensions*, ed. Kinzig, A. P., Pacala, S. W., and Tilman, D. Princeton, NJ/Oxford: Princeton University Press, pp. 294–313.

Lawton, J. H. (1994). What do species do in ecosystems? *Oikos*, **71**, 367–74.

Lawton, J. H. and Brown, V. K. (1994). Redundancy in ecosystems. In *Biodiversity and Ecosystem Function*, ed. Schulze, E.-D., and Mooney, H. A. Berlin: Springer, pp. 255–70.

Lawton, J. H., Naeem, S., Woodfin, R. M., Brown, V. K., Gange, A., Godfray, H. J. C., Heads, P. A., Lawler, S., Magda, D., Thomas, C. D., Thompson,

L. J., and Young, S. (1993). The ecotron: a controlled environmental facility for the investigation of population and ecosystem processes. *Philosophical Transactions of the Royal Society B*, **341**, 181–94.

Lees, K., Pitois, S., Scott, C., Frid, C., and Mackinson, S. (2006). Characterizing regime shifts in the marine environment. *Fish and Fisheries*, **7**, 104–27.

Leopold, A. (1949). The land ethic. In *A Sand County Almanach and Sketches Here and There*, ed. Leopold, A. New York, NY: Oxford University Press, pp. 201–26.

Leopold, A. S., Cain, S. A., Cottham, C. M., Gabrielson, I. M., and Kimball, T. L. (1963). Wildlife management in the national parks. *American Forests*, **63**, 32–5, 61–3.

Leps, J. (2004). What do the biodiversity experiments tell us about consequences of plant species loss in the real world? *Basic and Applied Ecology*, **5**, 529–34.

Leßmann, D. and Nixdorf, B. (2002). Probleme der Umsetzung der EU-Wasserrahmenrichtlinie bei Tagebauseen. *BTU Cottbus, Aktuelle Reihe*, **5**(2), 147–59.

Levin, S. A. (1998). Ecosystems and the biosphere as complex adaptive systems. *Ecosystems*, **1**, 431–6.

Levin, S. A. (2005). Self-organization and the emergence of complexity in ecological systems. *BioScience*, **55**, 1075–9.

Libralato, S., Torricelli, P., and Pranovi, F. (2006). Exergy as ecosystem indicator: an application to the recovery process of marine benthic communities. *Ecological Modelling*, **192**, 571–85.

Lindeman, R. L. (1942). The trophic-dynamic aspect of ecology. *Ecology*, **23**, 399–417.

Lizarralde, M. (1993). Current status of the introduced beaver (*Castor canadensis*) population in Tierra del Fuego, Argentina. *Ambio*, **22**, 351–8.

Longino, H. E. (1983). Beyond 'bad science': skeptical reflections on the value-freedom of scientific inquiry. *Science, Technology & Human Values*, **8**, 7–17.

Longino, H. E. (1990). *Science as Social Knowledge: Values and Objectivity in Scientific Inquiry*, Princeton, NJ: Princeton University Press.

Loreau, M. (2004). Does functional redundancy exist? *Oikos*, **104**, 606–11.

Loreau, M. (2008). Biodiversity and ecosystem functioning: the mystery of the deep sea. *Current Biology*, **18**, R126–8.

Loreau, M., Naeem, S., and Inchausti, P. (eds) (2002a). *Biodiversity and Ecosystem Functioning: Synthesis and Perspectives*, Oxford: Oxford University Press.

Loreau, M., Downing, A. L., Emmerson, M. C., Gonzalez, A., Hughes, J., Inchausti, P., Joshi, J., Norberg, J., and Sala, O. E. (2002b). A new look at the relationship between diversity and stability. In *Biodiversity and Ecosystem Functioning: Synthesis and Perspectives*, ed. Loreau, M., Naeem, S., and Inchausti, P. Oxford: Oxford University Press, pp. 79–91.

Loucks, O. L. (2000). Patterns of forest integrity in the eastern United States and Canada: measuring loss and recovery. In *Ecological Integrity: Integrating Environment, Conservation, and Health*, ed. Pimentel, D., Westra, L., and Noss, R. F. Washington, DC: Island Press, pp. 177–90.

Lovejoy, A. O. (1936 [1957]). *The Great Chain of Being: A Study of the History of an Idea*, Cambridge, MA: Harvard University Press.

Lovelock, J. (1979). *Gaia: A New Look on Earth*, Oxford: Oxford University Press.

Lovelock, J. (1989). Geophysiology, the science of Gaia. *Reviews in Geophysics*, **27**, 215–22.

Luck, G. W., Daily, G. C., and Ehrlich, P. R. (2003). Population diversity and ecosystem services. *Trends in Ecology & Evolution*, **18**, 331–6.

Luck, G. W., Harrington, R., Harrison, P., Kremen, C., Berry, P. M., Bugter, R., Dawson, T. P., de Bello, F., Diaz, S., Feld, C. K., Haslett, J. R., Hering, D., Kontogianni, A., Lavorel, S., Rounsevell, M., Samways, M. J., Sandin, L., Settele, J., Sykes, M. T., Van Den Hove, S., Vandewalle, M., and Zobel, M. (2009). Quantifying the contribution of organisms to the provision of ecosystem services. *Bioscience*, **59**, 223–35.

Ludovisi, A., Pandolfi, P., and Taticchi, M. I. (2005). The strategy of ecosystem development: specific dissipation as an indicator of ecosystem maturity. *Journal of Theoretical Biology*, **235**, 33–43.

Lyche Solheim, A. (ed.) (2005). *Reference Conditions of European Lakes: Indicators for the Water Framework Directive. Assessment of Reference Conditions*. REBECCA Project, Deliverable 7. [online] URL: http://www.rbm-toolbox.net/rebecca.

Lyche Solheim, A., Rekolainen, S., Moe, S. J., Carvalho, L., Phillips, G., Ptacnik, R., Penning, W. E., Toth, L. G., O'Toole, C., Schartau, A.-K. L., and Hesthagen, T. (2008). Ecological threshold responses in European lakes and their applicability for the Water Framework Directive (WFD) implementation: synthesis of lakes results from the REBECCA project. *Aquatic Ecology*, **42**, 317–34.

MacArthur, R. H. (1955). Fluctuations of animal populations, and a measure of community stability. *Ecology*, **36**, 533–6.

Mageau, M. T., Costanza, R., and Ulanowicz, R. E. (1995). The development and initial testing of a quantitative assessment of ecosystem health. *Ecosystem Health*, **1**, 201–13.

Magurran, A. E. (1988). *Ecological Diversity and Its Measurement*, Princeton, NJ: Princeton University Press.

Magurran, A. E. (2004). *Measuring Biological Diversity*, Oxford: Blackwell.

Mahner, M. and Bunge, M. (1997). *Foundations of Biophilosophy*, Berlin/Heidelberg/New York, NY: Springer.

Malone, C. R. (2000). Ecosystem management policies in state government of the USA. *Landscape and Urban Planning*, **48**, 57–64.

Marquard, E., Weigelt, A., Temperton, V. M., Roscher, C., Schumacher, J., Buchmann, N., Fischer, M., Weisser, W. W., and Schmid, B. (2009). Plant species richness and functional composition drive overyielding in a six-year grassland experiment. *Ecology*, **90**, 3290–302.

Martinez, N. D. (1996). Defining and measuring functional aspects of biodiversity. In *Biodiversity: A Biology of Numbers and Difference*, ed. Gaston, K. J. Oxford: Blackwell, pp. 114–48.

Martínez Pastur, G., Lencinas, M. V., Escobar, J., Quiroga, P., Malmierca, L., and Lizarralde, M. (2006). Understorey succession in *Nothofagus* forests in Tierra del Fuego (Argentina) affected by *Castor canadensis*. *Applied Vegetation Science*, **9**, 143–54.

Martinic, M. (2002). *Brief History of the Land of Magellan*, Punta Arenas: Ediciones de la Universidad de Magallanes.

Mason, N. W. H., Mouillot, D., Lee, W. G., and Wilson, J. B. (2005). Functional richness, functional evenness and functional divergence: the primary components of functional diversity. *Oikos*, **111**, 112–18.

May, R. M. (1974). *Stability and Complexity in Model-Ecosystems*, Princeton, NJ: Princeton University Press.

Mayr, E. (1988a). The multiple meanings of teleological. In *Towards a New Philosophy of Biology: Observations of an Evolutionist*, ed. Mayr, E. Cambridge, MA: Belknap Press of Harvard University Press, pp. 38–66.

Mayr, E. (1988b). The species category. In *Towards a New Philosophy of Biology: Observations of an Evolutionist*, ed. Mayr, E. Cambridge, MA: Belknap Press of Harvard University Press, pp. 315–34.

McCann, K. S. (2000). The diversity–stability debate. *Nature*, **405**, 228–33.

McCauley, D. J. (2006). Selling out on nature. *Nature*, **443**, 27–8.

McClelland, B. R. (1968). The ecosystem: a unifying concept for the management of natural areas in the national park system. MS Thesis, Colorado State University.

McConnell, F. (1996). *The Biodiversity Convention: A Negotiating History*, London: Kluwer Law International.

McCoy, E. D. (1996). Advocacy as part of conservation biology. *Conservation Biology*, **10**, 919–20.

McIntosh, R. P. (1998). The myth of community as organism. *Perspectives in Biology and Medicine*, **41**, 426–38.

McLauglin, P. (2001). *What Functions Explain: Functional Explanation and Self-reproducing Systems*, Cambridge: Cambridge University Press.

McLaughlin, P. (2005). Funktion. In *Philosophie der Biologie*, ed. Krohs, U. and Töpfer, G. Frankfurt am Main: Suhrkamp, pp. 19–35.

Meagher, M. and Houston, D. B. (1998). *Yellowstone and the Biology of Time*, Norman, OK: University of Oklahoma Press.

Meine, C. and Meffe, G. K. (1996). Conservation values, conservation science: a healthy tension. *Conservation Biology*, **10**, 916–17.

Meyer, J. L. (1997). Stream health: incorporating the human dimension to advance stream ecology. *Journal of the North American Benthological Society*, **16**, 439–47.

Miles, J. (1987). Vegetation succession: past and present concepts. In *Colonization, Succession and Stability*, ed. Gray, A. J., Crawley, M. J., and Edwards, P. J. Oxford: Blackwell, pp. 1–29.

Millennium Ecosystem Assessment (2003). *Ecosystems and Human Well-being: A Framework for Assessment*, Washington, DC: Island Press.

Millennium Ecosystem Assessment (2005). *Ecosystems and Human Well-being: Synthesis*, Washington, DC: Island Press.

Mitsch, W. J. and Jørgensen, S. E. (2003). Ecological engineering: a field whose time has come. *Ecological Engineering*, **20**, 363–77.

Mittermeier, R. A., Mittermeier, C. G., Brooks, T. M., Pilgrim, J. D., Konstant, W. R., Da Fonseca, G. A. B., and Kormos, C. (2003). Wilderness and biodiversity conservation. *Proceedings of the National Academy of Sciences of the United States of America*, **100**, 10309–13.

Moog, O. and Chovanec, A. (2000). Assessing the ecological integrity of rivers: walking the line among ecological, political and administrative interests. *Hydrobiologia*, **422–3**, 99–109.

Mooney, H. A. and Ehrlich, P. R. (1997). Ecosystem services: a fragmentary history. In *Nature's Service: Societal Dependence on Natural Ecosystems*, ed. Daily, G. C. Washington, DC: Island Press, pp. 11–19.

Moore, D. M. (1983). *Flora of Tierra del Fuego*, Oswestry: Anthony Nelson.

Moore, W. E. (1978). Functionalism. In *A History of Sociological Analysis*, ed. Bottomore, T. B. and Nisbet, R. A. New York, NY: Basic Books, pp. 321–61.

Moorman, M. C., Eggleston, D. B., Anderson, C. B., Mansilla, A., and Szejner, P. (2009). Implications of beaver *Castor canadensis* and trout introductions on native fish in the Cape Horn Biosphere Reserve, Chile. *Transactions of the American Fisheries Society*, **138**, 306–313.

Moss, B. (2007). Shallow lakes, the Water Framework Directive and life: What should it all be about? *Hydrobiologia*, **584**, 381–94.

Moss, B. (2008). The Water Framework Directive: total environment or political compromise? *Science of the Total Environment*, **400**, 32–41.

Moss, B., Stephen, D., Alvarez, C., Becares, E., Van de Bund, W., Collings, S. E., Van Donk, E., de Eyto, E., Feldmann, T., Fernández-Aláez, C., Fernández-Aláez, M., Franken, R. J. M., García-Criado, F., Gross, E. M., Gyllström, M., Hansson, L.-A., Irvine, K., Järvalt, A., Jensen, J.-P., Jeppesen, E., Kairesalo, T., Kornijów, R., Krause, T., Künnap, H., Laas, A., Lill, E., Lorens, B., Luup, H., Miracle, M. R., Nõges, P., Nõges, T., Nykänen, M., Ott, I., Peczula, W., Peeters, E. T. H. M., Phillips, G., Romo, S., Russell, V., Salujäe, J., Scheffer, M., Siewertsen, K., Smal, H., Tesch, C., Timm, H., Tuvikene, L., Tonno, I., Virro, T., Vicente, E., and Wilson, D. (2003). The determination of ecological status in shallow lakes: a tested system (ECOFRAME) for implementation of the European Water Framework Directive. *Aquatic Conservation: Marine and Freshwater Ecosystems*, **13**, 507–49.

Moulton, T. P. (1999). Biodiversity and ecosystem functioning in conservation of rivers and streams. *Aquatic Conservation: Marine and Freshwater Ecosystems*, **9**, 573–8.

Mulder, C. P. H., Bazeley-White, E., Dimitrakopoulos, P. G., Hector, A., Scherer-Lorenzen, M., and Schmid, B. (2004). Species evenness and productivity in experimental plant communities. *Oikos*, **107**, 50–63.

Müller, F. and Jørgensen, S. E. (2000). Ecological orientors: a path to environmental applications of ecosystem theories. In *Handbook of Ecosystem Theories and Management*, ed. Jørgensen, S. E. and Müller, F. Boca Raton, FL: Lewis Publishers, pp. 561–75.

Mumby, P. J., Hastings, A., and Edwards, H. J. (2007). Thresholds and the resilience of Caribbean coral reefs. *Nature*, **450**, 98–101.

Münch, R. (2003). Funktionalismus: Geshichte und Zukunftsperspektive einer Theorietradition. In *Soziologischer Funktionalismus. Zur Methodologie einer Theorietradition*, ed. Jetzkowitz, J. and Stark, C. Opladen: Leske & Budrich, pp. 17–56.

Nabakov, P. and Loendorf, L. (2002). *American Indians and Yellowstone National Park: A Documentary Overview*, Yellowstone National Park, WY: National Park Service, Yellowstone Center for Resources.

Naeem, S. (1998). Species redundancy and ecosystem reliability. *Conservation Biology*, **12**, 39–45.

Naeem, S. (2000). Reply to Wardle *et al. Bulletin of the Ecological Society of America*, **81**, 241–6.

Naeem, S. (2002). Ecosystem consequences of biodiversity loss: the evolution of a paradigm. *Ecology*, **83**, 1537–52.

Naeem, S. (2006). Biodiversity and ecosystem functioning in restored ecosystems: extracting principles for a synthetic perspective. In *Foundations of Restoration Ecology*, ed. Falk, D. A., Palmer, M. A., and Zedler, J. B. Washington, DC: Island Press, pp. 210–37.

Naeem, S. and Wright, J. P. (2003). Disentangling biodiversity effects on ecosystem functioning: deriving solutions to a seemingly insurmountable problem. *Ecology Letters*, **6**, 567–79.

Naeem, S., Loreau, M., and Inchausti, P. (2002). Biodiversity and ecosystem functioning: the emergence of a synthetic ecological framework. In *Biodiversity and Ecosystem Functioning: Synthesis and Perspectives*, ed. Loreau, M., Naeem, S., and Inchausti, P. Oxford: Oxford University Press, pp. 3–11.

Naeem, S., Bunker, D. E., Hector, A., Loreau, M., and Perrings, C. (eds) (2009a). *Biodiversity, Ecosystem Functioning, & Human Wellbeing: An Ecological and Economic Perspective*, Oxford: Oxford University Press.

Naeem, S., Bunker, D. E., Hector, A., Loreau, M., and Perrings, C. (2009b). Introduction: the ecological and social implications of changing biodiversity: an overview of a decade of biodiversity and ecosystem functioning research. In *Biodiversity, Ecosystem Functioning, & Human Wellbeing: An Ecological and Economic Perspective*, ed. Naeem, S., Bunker, D. E., Hector, A., Loreau, M., and Perrings, C. Oxford: Oxford University Press, pp. 3–13.

Naess, A. (1973). The shallow and the deep, long-range ecology movement: a summary. *Inquiry*, **16**, 95–100.

Nagel, E. (1961). *The Structure of Science: Problems in the Logic of Scientific Explanation*, New York, NY: Harcourt, Brace & World Inc.

Nagel, E. (1977). Functional explanations in biology. *Journal of Philosophy*, **74**, 280–301.

Naiman, R. J., Johnston, C. A., and Kelley, J. C. (1988). Alteration of North-American streams by beaver. *Bioscience*, **38**, 753–62.

Nash, R. (1982). *Wilderness and the American Mind*, New Haven, CT: Yale University Press.

National Park Service (1968). *Administrative Policies for Natural Areas of the National Park System*, Washington, DC: US Government Printing Office.

National Park Service (1978). *Management Policies*, Washington, DC: US Government Printing Office.

National Park Service (1988). *Management Policies*, Washington, DC: US Government Printing Office.

National Park Service (2006). *Management Policies 2006*, Washington, DC: US Government Printing Office.

National Research Council (2002). *Ecological Dynamics on Yellowstone's Northern Range*, Washington, DC: National Academy Press.

Nelson, M. P. and Callicott, J. B. (eds) (2008). *The Wilderness Debate Rages On*, Athens, GA: University of Georgia Press.

Neßhöver, C. (2005). The role of plant functional diversity in Central European grassland systems for ecosystem functioning. *Bayreuther Forum Ökologie*, **110**, 1–278.

Niemann, E. (1977). Eine Methode zur Erarbeitung der Funktionsleistungsgrade von Landschaftselementen. *Archiv für Naturschutz und Landschaftsforschung*, **17**, 119–57.

Nixdorf, B., Mischke, U., and Lessmann, D. (1998). Chrysophytes and chlamydomonads: pioneer colonists in extremely acidic mining lakes (pH < 3) in Lusatia (Germany). *Hydrobiologia*, **370**, 315–27.

Nixdorf, B., Lessmann, D., and Deneke, R. (2005). Mining lakes in a disturbed landscape: application of the EC Water Framework Directive and future management strategies. *Ecological Engineering*, **24**, 67–73.

Noble, J. R. and Slatyer, R. O. (1980). The use of vital attributes to predict successional changes in plant communities subject to recurrent disturbances. *Vegetatio*, **43**, 5–21.

Nõges, P., Van de Bund, W., Cardoso, A. C., Solimini, A. G., and Heiskanen, A.-S. (2009). Assessment of the ecological status of European surface waters: a work in progress. *Hydrobiologia*, **633**, 197–211.

Norris, R. H. and Thoms, M. C. (1999). What is river health? *Freshwater Biology*, **41**, 197–209.

Norton, B. G. (1992). A new paradigm for environmental management. In *Ecosystem Health: New Goals for Environmental Management*, ed. Costanza, R., Norton, B. G., and Haskell, B. D. Washington, DC: Island Press, pp. 23–41. ★★★★

Nyström, M. (2006). Redundancy and response diversity of functional groups: implications for the resilience of coral reefs. *Ambio*, **35**, 30–5.

Nyström, M., Folke, C., and Moberg, F. (2000). Coral reef disturbances and resilience in a human-dominated environment. *Trends in Ecology & Evolution*, **15**, 413–17.

Nyström, M., Graham, N. A. J., Lokrantz, J., and Norström, A. V. (2008). Capturing the cornerstones of coral reef resilience: linking theory to practice. *Coral Reefs*, **27**, 795–809.

O'Neill, R. V. (1976). Ecosystem persistence and heterotrophic regulation. *Ecology*, **57**, 1244–53.

O'Neill, R. V., Ausmus, B. S., Jackson, D. R., Vanhook, R. I., Vanvoris, P., Washburne, C., and Watson, A. P. (1977). Monitoring terrestrial ecosystems by analysis of nutrient export. *Water Air and Soil Pollution*, **8**, 271–7.

O'Neill, R. V., Deangelis, D. L., Waide, J. B., and Allen, T. F. H. (1986). *A Hierarchical Concept of Ecosystems*, Princeton, NJ: Princeton University Press.

Odum, E. P. (1985). Trends expected in stressed ecosystems. *Bioscience*, **35**, 419–22.

Odum, H. T. (1971). *Environment, Power and Society*, New York, NY: Wiley Interscience.

Odum, H. T. (1983). *Systems Ecology: An Introduction*, New York, NY: Wiley.

Olenin, S., Minchin, D., and Daunys, D. (2007). Assessment of biopollution in aquatic ecosystems. *Marine Pollution Bulletin*, **55**, 379–94.

Orians, G. H. (1975). Diversity, stability and maturity in natural ecosystems. In *Unifying Concepts in Ecology*, ed. Dobben, W. H. V. The Hague: Jungk, pp. 139–50.

Ott, K. (2008). A modest proposal to proceed in order to solve the problem of inherent moral value in nature. In *Reconciling Human Existence with Ecological Integrity*, ed. Westra, L., Bosselmann, K., and Westra, R. London: Earthscan, pp. 39–59.

Palmer, M., Bernhardt, E., Chornesky, E., Collins, S., Dobson, A., Duke, C., Gold, B., Jacobson, R., Kingsland, S., Kranz, R., Mappin, M., Martinez, M. L., Micheli, F., Morse, J., Pace, M., Pascual, M., Palumbi, S., Reichman, O. J., Simons, A., Townsend, A., and Turner, M. (2004). Ecology for a crowded planet. *Science*, **304**, 1251–2.

Palumbi, S. R., Sandifer, P. A., Allan, J. D., Beck, M. W., Fautin, D. G., Fogarty, M. J., Halpern, B. S., Incze, L. S., Leong, J.-A., Norse, E., Stachowicz, J. J., and Wall, D. H. (2009). Managing for ocean biodiversity to sustain marine ecosystem services. *Frontiers in Ecology and the Environment*, **7**, 204–11.

Parker, I. M., Simberloff, D., Lonsdale, W. M., Goodell, K., Wonham, M., Kareiva, P. M., Williamson, M. H., Von Holle, B., Byers, J. E., and Goldwasser, L. (1999). Impact: toward a framework for understanding the ecological effects of invaders. *Biological Invasions*, **1**, 3–19.

Parris, T. M. and Kates, R. W. (2003). Characterizing and measuring sustainable development. *Annual Review of Environment and Resources*, **28**, 559–86.

Petchey, O. L. and Gaston, K. J. (2006). Functional diversity: back to basics and looking forward. *Ecology Letters*, **9**, 741–58.

Petersen, C. G. J. (1913). Valuation of the sea: II. The animal communities of the sea-bottom and their importance for marine zoogeography. *Report of the Danish Biological Station to the Board of Agriculture*, **21**, 1–44.

Phillips, J. (1935). Succession, development, the climax, and the complex organism: an analysis of concepts. Part III: the complex organism. Conclusions. *Journal of Ecology*, **23**, 488–508.

Philpott, S. M., Soong, O., Lowenstein, J. H., Luz Polido, A., Tobar Lopez, D., Flynn, D. F. B., and Declerck, F. (2009). Functional richness and ecosystem services: bird predation on arthropods in tropical agroecosystems. *Ecological Applications*, **19**, 1858–67.

Pickering, A. (1995). *The Mangle of Practice: Time, Agency, and Science*, Chicago, IL/London: University of Chicago Press.

Pickett, S. T. A. and Cadenasso, M. L. (2002). The ecosystem as a multidimensional concept: meaning, model, and metaphor. *Ecosystems*, **5**, 1–10.

Pickett, S. T. A. and Ostfeld, R. S. (1995). The shifting paradigm in ecology. In *A New Century for Natural Resources Management*, ed. Knight, R. L. and Bates, S. F. Washington, DC: Island Press, pp. 261–78.

Pickett, S. T. A., Collins, S. L., and Armesto, J. J. (1987). A hierarchical consideration of causes and mechanisms of succession. *Vegetatio*, **69**, 109–14.

Pickett, S. T. A., Kolasa, J., and Jones, C. G. (1994). *Ecological Understanding*, San Diego, CA: Academic Press.

Pielke, Jr., R. (2007). *The Honest Broker: Making Sense of Science in Policy and Politics*, Cambridge: Cambridge University Press.

Pimentel, D., Westra, L., and Noss, R. F. (eds) (2000). *Ecological Integrity: Integrating Environment, Conservation, and Health*, Washington, DC: Island Press.

Pimm, S. L. (1984). The complexity and stability of ecosystems. *Nature*, **307**, 321–6.

Platt, J. (1969). Theorems on boundaries in hierarchical systems. In *Hierarchical Structures*, ed. Whyte, L. L., Wilson, A. G., and Wilson, D. New York, NY: Elsevier, pp. 201–13.

Potthast, T. (2000a). Bioethics and epistemic–moral hybrids: perspectives from the history of science. *Biomedical Ethics*, **5**, 20–3.

Potthast, T. (2000b). Funktionssicherung und/oder Aufbruch ins Ungewisse? Anmerkungen zum Prozeßschutz. In *Funktionsbegriff und Unsicherheit in der Ökologie*, ed. Jax, K. Frankfurt am Main: Peter Lang, pp. 65–81.

Potthast, T. (2005). Umweltforschung und das Problem epistemisch-moralischer Hybride: Ein Kommentar zur Rhetorik, Programmatik und Theorie interdisziplinärer Forschung. In *Wissenschaftsphilosophie interdisziplinärer Umweltforschung*, ed. Baumgärtner, S. and Becker, C. Marburg: Metropolis, pp. 87–100.

Prach, K. and Hobbs, R. J. (2008). Spontaneous succession versus technical reclamation in the restoration of disturbed sites. *Restoration Ecology*, **16**, 363–6.

Prach, K., Bartha, S., Joyce, C. B., Pyšek, P., Van Diggelen, R., and Wiegleb, G. (2001). The role of spontaneous vegetation succession in ecosystem restoration: a perspective. *Applied Vegetation Science*, **4**, 111–14.

Pranger, R. (1990). Towards a pluralistic concept of function: function statements in biology. *Acta Biotheoretica*, **38**, 63–71.

Pritchard, J. A. (1999). *Preserving Yellowstone's Natural Conditions: Science and the Perception of Nature*, Lincoln, NE: University of Nebraska Press.

Proctor, J. D. (1998). The social construction of nature: relativist accusations, pragmatist and critical realist responses. *Annals of the Association of American Geographers*, **88**, 352–76.

Pyšek, P., Richardson, D. M., Rejmánek, M., Webster, G. L., Williamson, M., and Kirscher, J. (2004). Alien plants in checklists and floras: towards better communication between taxonomists and ecologists. *Taxon*, **53**, 131–43.

Radder, H. (1992). Normative reflexions on constructivist approaches to science and technology. *Social Studies of Science*, **22**, 141–73.

Rapport, D. J. (1989). What constitutes ecosystem health? *Perspectives in Biology and Medicine*, **33**, 120–32.

Rapport, D. J. (1992). What is clinical ecology? In *Ecosystem Health: New Goals for Environmental Management*, ed. Costanza, R., Norton, B. G., and Haskell, B. D. Washington, DC: Island Press, pp. 144–56.

Rapport, D. J. (1995). Ecosystem health: exploring the territory. *Ecosystem Health*, **1**, 5–13.

Rapport, D. J., Regier, H. A., and Hutchinson, T. C. (1985). Ecosystem behavior under stress. *American Naturalist*, **125**, 617–40.

Raunkiaer, C. (1934). The statistics of life-forms as a basis for biological plant geography. In *The Life-forms of Plants and Statistical Plant Geography*, ed. Raunkiaer, C. Oxford: Clarendon Press, pp. 111–47.

Ravnborg, H. M. and Westermann, O. (2002). Understanding interdependencies: stakeholder identification and negotiation for collective natural resource management. *Agricultural Systems*, **73**, 41–56.

Rebertus, A. J. and Veblen, T. T. (1993). Structure and tree-fall gap dynamics of old-growth *Nothofagus* forests in Tierra del Fuego, Argentina. *Journal of Vegetation Science*, **4**, 641–54.

Reed, M. S. (2008). Stakeholder participation for environmental management: a literature review. *Biological Conservation*, **141**, 2417–31.

Regier, H. A. (1993). The notion of natural and cultural integrity. In *Ecological Integrity and the Management of Ecosystems*, ed. Woodley, S., Kay, J., and Francis, G. Delray Beach, FL: St. Lucie Press, pp. 3–18.

Reid, N. J. (1968). Ecosystem management in the national parks. In *Transactions of the Thirty-Third North American Wildlife and Natural Resources Conference, March 11, 12, 13, 1968*, Washington, DC: Wildlife Management Institute, pp. 160–9.

Reiss, J., Bridle, J. R., Montoya, J. M., and Woodward, G. (2009). Emerging horizons in biodiversity and ecosystem functioning research. *Trends in Ecology & Evolution*, **24**, 505–14.

Ridder, B. (2007). The naturalness versus wildness debate: ambiguity, inconsistency, and unattainable objectivity. *Restoration Ecology*, **15**, 8–12.

Ridder, B. (2008). Questioning the ecosystem services argument for biodiversity conservation. *Biodiversity and Conservation*, **17**, 781–90.

Ripple, W. J. and Beschta, R. L. (2004). Wolves and the ecology of fear: can predation risk structure ecosystems? *BioScience*, **54**, 755–66.

Ripple, W. J. and Beschta, R. L. (2007). Restoring Yellowstone's aspen with wolves. *Biological Conservation*, **138**, 514–19.

Rolston, H. I. (1990). Biology and philosophy in Yellowstone. *Biology and Philosophy*, **5**, 241–58.

Rolston, H. I. (1994). Value in nature and the nature of value. In *Philosophy and the Natural Environment*, ed. Attfield, R. and Belsey, A. Cambridge: Cambridge University Press, pp. 13–30.

Roscher, C., Schumacher, J., Baade, J., Wilcke, W., Gleixner, G., Weisser, W. W., Schmid, B., and Schulze, E.-D. (2004). The role of biodiversity for element cycling and trophic interactions: an experimental approach in a grassland community. *Basic and Applied Ecology*, **5**, 107–21.

Rosenberg, A. (1985). *The Structure of Biological Science*, Cambridge: Cambridge University Press.

Rosenfeld, J. S. (2002). Logical fallacies in the assessment of functional redundancy. *Conservation Biology*, **16**, 837–9.

Rozzi, R., Massardo, F., Berghöfer, A., Anderson, C. B., Mansilla, A., Plana, J., Berghöfer, U., Araya, P., and Barros, E. (2006). *Cape Horn Biosphere Reserve: Nomination Document of the Cape Horn Archipelago Territory into the World Biosphere Reserve Network*. MAB Programme: UNESCO, Punta Arenas: Edicones Universidad de Magallanes.

Rozzi, R., Armesto, J. J., Goffinet, B., Buck, W., Massardo, F., Silander, J., Arroyo, M. T. K., Russell, S., Anderson, C. B., Cavieres, L. A., and Callicott, J. B. (2008). Changing lenses to assess biodiversity: patterns of species richness in sub-Antarctic plants and implications for global conservation. *Frontiers in Ecology and the Environment*, **6**, 131–7.

Ruesink, J. L. and Srivastava, D. S. (2001). Numerical and per capita responses to species loss: mechanisms maintaining ecosystem function in a community of stream insect detritivores. *Oikos*, **93**, 221–34.

Ruse, M. (1987). Biological species: natural kinds, individuals, or what? *British Journal for the Philosophy of Science*, **38**, 225–42.

Ruse, M. (1989). Teleology in biology: is it a cause for concern? *Trends in Ecology & Evolution*, **4**, 51–4.

Sagoff, M. (2003). The plaza and the pendulum: two concepts of ecological science. *Biology and Philosophy*, **18**, 529–52.

Sanders, R. W. (1991). Mixotrophic protists in marine and freshwater ecosystems. *Journal of Protozoology*, **38**, 76–81.

Sandin, L. and Solimini, A. G. (2009). Freshwater ecosystem structure–function relationships: from theory to application. *Freshwater Biology*, **54**, 2017–24.

Sasaki, N. and Putz, F. E. (2009). Critical need for new definitions of 'forest' and 'forest degradation' in global climate change agreements. *Conservation Letters*, **2**, 226–32.

Schabel, H. G. and Palmer, S. L. (1999). The Dauerwald: its role in the restoration of natural forests. *Journal of Forestry*, **97**, 20–5.

Scharf, B. W. and Björk, S. (eds) (1992). *Limnology of Eifel Maar Lake: Ergebnisse der Limnologie, Heft 38*, Stuttgart: Schweizerbartsche Verlagsbuchhandlung.

Scheffer, M. and Carpenter, S. R. (2003). Catastrophic regime shifts in ecosystems: linking theory to observation. *Trends in Ecology & Evolution*, **18**, 648–56.

Scheffer, M., Hosper, S. H., Meijer, M. L., Moss, B., and Jeppesen, E. (1993). Alternative equilibria in shallow lakes. *Trends in Ecology & Evolution*, **8**, 275–9.

Scherzinger, W. (1990). Das Dynamik-Konzept im flächenhaften Naturschutz: Zieldiskussion am Beispiel der Nationalpark-Idee. *Natur und Landschaft*, **65**, 292–8.

Schläpfer, F. and Schmidt, B. (1999). Ecosystem effects of biodiversity: a classification of hypotheses and exploration of empirical results. *Ecological Applications*, **9**, 893–912.

Schmid, B. and Hector, A. (2004). The value of biodiversity experiments. *Basic and Applied Ecology*, **5**, 535–42.

Schmid, B., Hector, A., Huston, M. A., Inchausti, P., Nijs, I., Leadley, P. W., and Tilman, D. (2002). The design and analysis of biodiversity experiments. In *Biodiversity and Ecosystem Functioning: Synthesis and Perspectives*, ed. Loreau, M., Naeem, S., and Inchausti, P. Oxford: Oxford University Press, pp. 61–75.

Schmidtz, D. and Willott, E. (eds) (2002). *Environmental Ethics: What Really Works*, New York, NY: Oxford University Press.

Schröder, A., Persson, L., and de Roos, A. M. (2005). Direct empirical evidence for alternative stable states: a review. *Oikos*, **110**, 3–19.

Schullery, P. (1997). *Searching for Yellowstone: Ecology and Wonder in the Last Wilderness*, Boston, MA/New York, NY: Houghton Mifflin.

Schultz, A. M. (1967). The ecosystem as a conceptual tool in the management of natural resources. In *Natural Resources: Quality and Quantity*, ed. Ciriacy-Wantrup, S. V. and Parsons, J. J. Berkeley, CA: University of California Press, pp. 139–61.

Schulz, F. and Wiegleb, G. (2000). Development options of natural habitats in a post-mining landscape. *Land Degradation & Development*, **11**, 99–110.

Schulze, E.-D. and Mooney, H. A. (eds) (1993). *Biodiversity and Ecosystem Function*, Berlin: Springer.

Schwartz, M. W., Brigham, C. A., Hoeksema, J. D., Lyons, K. G., Mills, M. H., and Van Mantgem, P. J. (2000). Linking biodiversity to ecosystem function: implications for conservation ecology. *Oecologia*, **122**, 297–305.

Schwind, W. (1984). *Der Eifelwald im Laufe der Jahrhunderte*, Düren: Eifelverein.

Seastedt, T. R., Hobbs, R. J., and Suding, K. N. (2008). Management of novel ecosystems: are novel approaches required? *Frontiers in Ecology and the Environment*, **6**, 547–53.

Seddon, P. J., Armstrong, D. P., and Maloney, R. F. (2007). Combining the fields of reintroduction biology and restoration ecology. *Conservation Biology*, **21**, 1388–90.

Sellars, R. W. (1997). *Preserving Nature in the National Parks: A History*, New Haven, CT: Yale University Press.

Servicio Agrícola Y Ganadero (2003). *Programa Control de Fauna Dañina en la XII Regón*, Punta Arenas: Servicio Agrícola y Ganadero, XII Region.

Sheperd, G. (2004). *The Ecosystem Approach: Five Steps to Implementation*, Gland, Switzerland: IUCN.

Sheperd, G. (ed.) (2008). *The Ecosystem Approach: Learning from Experience*, Gland, Switzerland: IUCN.

Shrader-Frechette, K. (1996). Throwing out the bathwater of positivism, keeping the baby of objectivity: relativism and advocacy in conservation biology. *Conservation Biology*, **10**, 912–14.

Shrader-Frechette, K. S. and McCoy, E. D. (1993). *Method in Ecology: Strategies for Conservation*, Cambridge: Cambridge University Press.

Sielfeld, W. and Venegas, C. (1980). Poblamiento e impacto ambiente de *Castor canadensis* KUHL, en la Isla Navarino, Chile. *Anales del Institutio de la Patagonia*, **11**, 247–57.

Silow, E. A. and In-Hye, O. (2004). Aquatic ecosystem assessment using exergy. *Ecological Indicators*, **4**, 189–98.

Simberloff, D. (1998). Flagships, umbrellas, and keystones: is single-species management passé in the landscape era? *Conservation Biology*, **83**, 247–57.

Singer, F. J. (ed.) (1996). *Effects of Grazing by Wild Ungulates in Yellowstone National Park: Technical Report NPS/NRYELL/NRTR/96–01*, Denver, CO: USDI/NPS.

Skewes, O., González, F., Rubilar, L., Quezada, M., Olave, R., Vargas, V., and Ávila, A. (1999). *Investigacón, Aprovechamiento y Control del Castor en Islas Tierra del Fuego y Navarino*, Punta Arenas: Servicio Agrícola y Ganadero, XII Region.

Skewes, O., González, F., Olave, R., Ávila, A., Vargas, V., Paulsen, P., and König, H. E. (2006). Abundance and distribution of American beaver, *Castor canadensis* (Kuhl 1820), in Tierra del Fuego and Navarino Islands, Chile. *European Journal of Wildlife Research*, **52**, 292–6.

Smith, M. D. and Knapp, A. K. (2003). Dominant species maintain ecosystem function with non-random species loss. *Ecology Letters*, **6**, 509–17.

Smith, T. M., Shugart, H. H., and Woodward, F. I. (1997). *Plant Functional Types*, Cambridge: Cambridge University Press.

Smith, V. H. and Schindler, D. W. (2009). Eutrophication science: where do we go from here? *Trends in Ecology & Evolution*, **24**, 201–7.

Society for Ecological Restoration (2004). *The SER International Primer on Ecological Restoration*, Tucson, AZ: SER International.

Solan, M., Cardinale, B. J., Downing, A. L., Engelhardt, K. A. M., Ruesink, J. L., and Srivastava, D. S. (2004). Extinction and ecosystem function in the marine benthos. *Science*, **306**, 1177–80.

Solimini, A. G., Ptacnik, R., and Cardoso, A. C. (2009). Towards holistic assessment of the functioning of ecosystems under the Water Framework Directive. *TrAC Trends in Analytical Chemistry*, **28**, 143–9.

Soulé, M. E. (1985). What is conservation biology? *BioScience*, **35**, 727–34.

Soulé, M. E. and Lease, G. (1995). Preface. In *Reinventing Nature? Responses to Postmodern Deconstruction*, ed. Soulé, M. E. and Lease, G. Washington, DC: Island Press, pp. xv–xvii.

Spehn, E. M., Hector, A., Joshi, J., Scherer-Lorenzen, M., Schmid, B., Bazeley-White, E., Beierkuhnlein, C., Caldeira, M. C., Diemer, M., Dimitrakopoulos, P. G., Finn, J. A., Freitas, H., Giller, P. S., Good, J., Harris, R., Hogberg, P., Huss-Danell, K., Jumpponen, A., Koricheva, J., Leadley, P. W., Loreau, M., Minns, A., Mulder, C. P. H., O'Donovan, G., Otway, S. J., Palmborg, C., Pereira, J. S., Pfisterer, A. B., Prinz, A., Read, D. J., Schulze, E. D., Siamantziouras, A. S. D., Terry, A. C., Troumbis, A. Y., Woodward, F. I., Yachi, S., and Lawton, J. H. (2005). Ecosystem effects of biodiversity manipulations in European grasslands. *Ecological Monographs*, **75**, 37–63.

Srivastava, D. S. and Vellend, M. (2005). Biodiversity–ecosystem function research: is it relevant to conservation? *Annual Review of Ecology Evolution and Systematics*, **36**, 267–94.

Stachowicz, J. J., Bruno, J. F., and Duffy, J. E. (2007). Understanding the effects of marine biodiversity on communities and ecosystems. *Annual Review of Ecology Evolution and Systematics*, **38**, 739–66.

Star, S. L. and Griesemer, J. R. (1989). Institutional ecology, translations and boundary objects: amateurs and professionals in Berkeley's Museum of Vertebrate Zoology, 1907–39. *Social Studies of Science*, **19**, 387–420.

Stegmüller, W. (1986). Wertfreiheit, Interessen und Objektivität: das Wertfreiheitspostulat von Max Weber. In *Rationale Rekonstruktion von Wissenschaft und ihrem Wandel*, ed. Stegmüller, W. Stuttgart: Reclam, pp. 177–203.

Sterelny, K. and Griffiths, P. E. (1999). *Sex and Death: An Introduction to Philosophy of Biology*, Chicago, IL: University of Chicago Press.

Steyaert, P. and Ollivier, G. (2007). The European Water Framework Directive: how ecological assumptions frame technical and social change. *Ecology and Society*, **12**. [online]: http://www.ecologyandsociety.org/vol12/iss1/art25.

Stoll-Kleemann, S. and Welp, M. (eds) (2006). *Stakeholder Dialogues in Natural Resources Management: Theory and Practice*, Berlin: Springer.

Sturm, K. (1993). Prozeßschutz: ein Konzept für naturschutzgerechte Waldwirtschaft. *Zeitschrift für Ökologie und Naturschutz*, **2**, 181–92.

Sukhdev, P. E. A. (2008). *The Economics of Ecosystems and Biodiversity (TEEB): An Interim Report*, Brussels: EU Commission.

Suter, G. W. I. (1993). A critique of ecosystem health concepts and indexes. *Environmental Toxicology and Chemistry*, **12**, 1533–9.

Swift, M. J., Izac, A. M. N., and Van Noordwijk, M. (2004). Biodiversity and ecosystem services in agricultural landscapes: are we asking the right questions? *Agriculture Ecosystems & Environment*, **104**, 113–34.

Takacs, D. (1996). *The Idea of Biodiversity: Philosophies of Paradise*, Baltimore, MD/London: Johns Hopkins University Press.

Tansley, A. G. (1935). The use and abuse of vegetational concepts and terms. *Ecology*, **16**, 284–307.

Taylor, P. J. (1988). Technocratic optimism, H.T. Odum, and the partial transformation of ecological metaphor after World War II. *Journal of the History of Biology*, **21**, 213–44.

Thiel, C. (1992). Funktion(alismus). In *Handlexikon zur Wissenschaftstheorie*, ed. Seiffert, H. and Radnitzky, G. München: dtv, pp. 86–8.

Thienemann, A. (1923). Die Gewässer Mitteleuropas: Eine hydrobiologische Charakterstik ihrer Haupttypen. In *Handbuch der Binnenfischerei Mitteleuropas*, Stuttgart: Schweizerbart'sche Verlagsbuchhandlung, pp. 1–84.

Thienemann, A. (1925). Der See als Lebenseinheit. *Naturwissenschaften*, **13**, 589–600.

Thompson, R. and Starzomski, B. M. (2007). What does biodiversity actually do? A review for managers and policy makers. *Biodiversity and Conservation*, **16**, 1359–78.

Tilman, D. (1997). Biodiversity and ecosystem functioning. In *Nature's Services: Societal Dependence on Natural Ecosystems*, ed. Daily, G. C. Washington, DC: Island Press, pp. 93–112.

Tilman, D., Wedin, D., and Knops, J. (1996). Productivity and sustainability influenced by biodiversity in grassland ecosystems. *Nature*, **379**, 718–20.

Tilman, D., Knops, J., Wedin, D., and Reich, P. (2001). Experimental and observational studies of diversity, productivity, and stability. In *The Functional Consequences of Biodiversity: Empirical Progress and Theoretical Extensions*, ed. Kinzig, A. P., Pacala, S. W., and Tilman, D. Princeton, NJ/Oxford: Princeton University Press, pp. 42–70.

Tischew, S. (ed.) (2004). *Renaturierung nach dem Braunkohleabbau*, Stuttgart: Teubner.

Töpfer, G. (2005). Teleologie. In *Philosophie der Biologie*, ed. Krohs, U. and Töpfer, G. Frankfurt: Suhrkamp, pp. 36–52.

Trepl, L. (1983). Ökologie: eine grüne Leitwissenschaft? Über Grenzen und Perspektiven einer modischen Disziplin. *Kursbuch*, **74**, 6–27.

Trepl, L. (1987). *Geschichte der Ökologie: Vom 17. Jahrhundert bis zur Gegenwart*, Frankfurt am Main: Athenäum.

Trepl, L. (1995). Die Diversitäts-Stabilitäts-Diskussion in der Ökologie. *Berichte der Bayerischen Akademie für Naturschutz und Landschaftspflege (ANL), Beiheft*, **12**, 35–49.

Tucker, R. W. (1979). The value decisions we know as science. *Advances in Nursing Science*, **1**, 1–12.

Tuhkanen, S. (1992). The climate of Tierra del Fuego from a vegetation geographical point of view and its ecoclimatic counterparts elsewhere. *Acta Botanica Fennica*, **145**, 1–64.

Tuhkanen, S., Kuokka, I., Hyvönen, J., Stenroos, S., and Nielmä, J. (1989). Tierra del Fuego as a target for biogeographical research in the past and present. *Anales del Institutio de la Patagonia*, **19**, 1–107.

Turner, M. G. (2008). Another perspective on Yellowstone's northern range. *BioScience*, **58**, 173–5.

Ulanowicz, R. E. (2000). Toward the measurement of ecological integrity. In *Ecological Integrity: Integrating Environment, Conservation, and Health*, ed. Pimentel, D., Westra, L., and Noss, R. F. Washington, DC: Island Press, pp. 99–113.

UNEP (2008). *UNEP Ecosystem Management Programme: An Ecosystem Approach. 10th Global Meeting on the Regional Seas.* Guayaquil. Ecuador, 25–27 November 2008, UNEP (DEPI)/RS.10/4. Available at: http://www.cbd.int/doc/external/unep/unep-eco-2008-11-en.pdf.

UNEP/CBD (2000). Ecosystem approach. In *Decisions Adopted by the Conference of the Parties to the Convention on Biological Diversity at its Fifth Meeting, Nairobi, 15–26 May 2000*, accessed 8 January 2009. Available at: http://www.cbd.int/doc/decisions/COP-05-dec-en.pdf, pp. 103–9.

UNESCO (2000). *Solving the Puzzle: The Ecosystem Approach and Biosphere Reserves*, Paris: UNESCO.

US Fish and Wildlife Service (1995). *Ecosystem Approach to Fish and Wildlife Conservation: Concept Document*, Washington, DC: US Fish and Wildlife Service. Available at: http://www.fws.gov/policy/npi95_03.pdf.

Valéry, L., Fritz, H., Lefeuvre, J.-C., and Simberloff, D. (2009). Ecosystem-level consequences of invasions by native species as a way to investigate relationships between evenness and ecosystem function. *Biological Invasions*, **11**, 609–17.

van der Maarel, E. (1997). *Biodiversity: From Babel to Biosphere Management*, Uppsala: Opulus Press.

van der Molen, D. T., Portielje, R., Boers, P. C. M., and Lijklema, L. (1998). Changes in sediment phosphorus as a result of eutrophication and oligotrophication in Lake Veluwe, the Netherlands. *Water Research*, **32**, 3281–8.

van der Steen, W. J. (1990). Concepts in biology: a survey of practical methodological principles. *Journal of Theoretical Biology*, **143**, 383–403.

van Langevelde, F., Van de Vijver, C. A. D. M., Kumar, L., Van de Koppel, J., de Ridder, N., Van Andel, J., Skidmore, A. K., Hearne, J. W., Stroosnijder, L., Bond, W. J., Prins, H. H., and Rietkerk, M. (2003). Effects of fire and herbivory on the stability of savanna ecosystems. *Ecology*, **84**, 337–50.

Villagrán de León, J. C. (2006). *Vulnerability: A Conceptual and Methodological Review*, Bonn: UNU-EHS.

Vitousek, P. M. (1990). Biological invasions and ecosystem processes: towards an integration of population biology and ecosystem studies. *Oikos*, **57**, 7–13.

Vitousek, P. M. and Hooper, D. U. (1993). Biological diversity and terrestrial biogeochemistry. In *Biodiversity and Ecosystem Function*, ed. Schulze, E.-D. and Mooney, H. A. Berlin: Springer, pp. 3–14.

Voigt, A. (2009). *Die Konstruktion der Natur: Ökologische Theorien und politische Philosphien der Vergesellschaftung*, Stuttgart: Franz Steiner Verlag.

Vos, W., Harms, B., and Stortelder, A. (1979). Einige Beispiele der Anwendung landschaftsökologischer Erkenntnisse in der Raumplanung in den Niederlanden. *Verhandlungen der Gesellschaft für Ökologie*, **7**, 85–99.

Wagner, F. H. (2006). *Yellowstone's Destabilized Ecosystem: Elk Effects, Science, and Policy Conflict*, Oxford: Oxford University Press.

Wagner, F. H. and Kay, C. E. (1993). 'Natural' or 'healthy' ecosystems: are US National Parks providing them? In *Humans as Components of Ecosystems: The Ecology of Subtle Human Effects and Populated Areas*, ed. McDonnell, J. M. and Pickett, S. T. A. New York, NY: Springer, pp. 257–70.

Wagner, F. H., Foresta, R., Gill, R. B., McCullough, D. R., Pelton, M. R., Porter, W. F., and Salwasser, H. (1995). *Wildlife Policies in the US National Parks*, Washington, DC: Island Press.

Walker, B. (1992). Biodiversity and redundancy. *Conservation Biology*, **6**, 18–23.

Walker, B. (1995). Conserving biological diversity through ecosystem resilience. *Conservation Biology*, **9**, 747–52.

Walker, B. and Salt, D. (2006). *Resilience Thinking: Sustaining Ecosystems and People in a Changing World*, Washington, DC: Island Press.

Walker, B., Carpenter, S., Anderies, J., Abel, N., Cumming, G., Janssen, M., Lebel, L., Norberg, J., Peterson, G. D., and Pritchard, R. (2002). Resilience management in social–ecological systems: a working hypothesis for a participatory approach. *Conservation Ecology*, **6**: 14 [online] http://www.consecol.org/vol6/iss1/art14.

Walker, B., Holling, C. S., Carpenter, S. R., and Kinzig, A. (2004). Resilience, adaptability and transformability in social–ecological systems. *Ecology and Society*, **9**(2), 5. [online] URL: http://www.ecologyandsociety.org/vol9/iss2/art5.

Wallace, K. J. (2007). Classification of ecosystem services: problems and solutions. *Biological Conservation*, **139**, 235–46.

Wardle, D. A., Huston, M. A., Grime, J. P., Berendse, F., Garnier, E., Lauenroth, W. K., Setälä, H., and Wilson, S. D. (2000). Biodiversity and ecosystem function: an issue in ecology. *Bulletin of the Ecological Society of America*, **81**, 235–9.

Warren, C. R. (2007). Perspectives on the 'alien' versus 'native' species debate: a critique of concepts, language and practice. *Progress in Human Geography*, **31**, 427–46.

Warzocha, J. (1995). Classification and structure of macrofaunal communities in the southern Baltic. *Archive of Fishery and Marine Research*, **42**, 225–37.

Weber, M. (1917 [1968]). Der Sinn der 'Wertfreiheit' der soziologischen und ökonomischen Wissenschaften. In *Gesammelte Aufsätze zur Wissenschaftslehre*, ed. Weber, M. Tübingen: J.C.B. Mohr, pp. 489–540.

Webler, T., Tuler, S., and Krueger, R. (2001). What is a good public participation process? Five perspectives from the public. *Environmental Management*, **27**, 435–50.

Weil, A. (2005). *Das Modell 'Organismus' in der Ökologie: Möglichkeiten und Grenzen der Beschreibung synökologischer Einheiten*, Frankfurt: Peter Lang.

Werner, R., Jax, K., and Böhmer, H. J. (2009). Vegetationsdynamik verlandeter Biberteiche auf der Insel Navarino (Feuerland-Archipel, Chile). *Tuexenia*, **29**, 277–96.

Westra, L. (1994). *An Environmental Proposal for Ethics: The Principle of Integrity*, Lanham, MD: Rowman & Littlefield.

Westra, L. (1998). The ethics of integrity. In *The Land Ethic: Meeting Human Needs for Land and its Resources*, ed. Society of American Foresters Bethesda, MD: Society of American Foresters, pp. 31–44.

Westra, L., Miller, P., Karr, J. R., Rees, W. E., and Ulanowicz, R. E. (2000). Ecological integrity and the aims of the Global Integrity Project. In *Ecological Integrity: Integrating Environment, Conservation, and Health*, ed. Pimentel, D., Westra, L., and Noss, R. F. Washington, DC: Island Press, pp. 19–41.

Whisenant, S. G. (1999). *Repairing Damaged Wildlands: A Process-oriented, Landscape-scale Approach*, Cambridge: Cambridge University Press.

White, P. S. and Jentsch, A. (2001). The search for generality in studies of disturbance and ecosystem dynamics. *Progress in Botany*, **62**, 399–450.

White, P. S. and Pickett, S. T. A. (1985). Natural disturbance and patch dynamics: an introduction. In *The Ecology of Natural Disturbance and Patch Dynamics*, ed. Pickett, S. T. A. and White, P. S. San Diego, CA: Academic Press, pp. 3–13.

Wicklum, D. and Davies, R. W. (1995). Ecosystem health and integrity? *Canadian Journal of Botany*, **73**, 997–1000.

Wiegleb, G. (1989). Explanation and prediction in vegetation science. *Vegetatio*, **83**, 17–34.

Wiegleb, G. (1996). Konzepte der Hierarchie-Theorie in der Ökologie. In *Systemtheorie in der Ökologie*, ed. Mathes, K., Breckling, B., and Ekschmitt, K. Landsberg: Ecomed, pp. 7–24.

Wiegleb, G. and Felinks, B. (2001). Primary succession in post-mining landscapes of Lower Lusatia: chance or necessity? *Ecological Engineering*, **17**, 199–217.

Wiegleb, G., Bröring, U., Mrzljak, J., and Schulz, F. (eds) (2000). *Naturschutz in Bergbaufolgelandschaften: Landschaftsanalyse und Leitbildentwicklung*, Heidelberg: Physica-Verlag.

Wilkinson, D. M. (1999). Gaia and natural selection. *Trends in Ecology and Evolution*, **14**, 256–7.

Williams, L. R. R. and Kapustka, L. A. (2000). Ecosystem vulnerability: a complex interface with technical components. *Environmental Toxicology and Chemistry*, **19**, 1055–8.

Wilsey, B. J. and Potvin, C. (2000). Biodiversity and ecosystem functioning: importance of species evenness in an old field. *Ecology*, **81**, 887–92.

Wilson, D. S. and Sober, E. (1989). Reviving the superorganism. *Journal of Theoretical Biology*, **136**, 337–56.

Wilson, E. O. (ed.) (1988). *Biodiversity*, Washington, DC: National Academy Press.

Wilson, J. B. (1999). Guilds, functional types and ecological groups. *Oikos*, **86**, 507–22.

Wilson, K., Pressey, R. L., Newton, A., Burgman, M., Possingham, H. P., and Weston, C. (2005). Measuring and incorporating vulnerability into conservation planning. *Environmental Management*, **35**, 527–43.

Winfree, R. and Kremen, C. (2009). Are ecosystem services stabilized by differences among species? A test using crop pollination. *Proceedings of the Royal Society B-Biological Sciences*, **276**, 229–37.

Wöllecke, J., Anders, K., Durka, W., Elmer, M., Wanner, M., and Wiegleb, G. (eds) (2007). *Landschaft im Wandel: Natürliche und Anthropogene Besiedlung der Niederlausitzer Bergbaufolgelandschaft*, Aachen: Shaker-Verlag.

Wollmann, K., Deneke, R., Nixdorf, B., and Packroff, G. (2000). Dynamics of planktonic food webs in three mining lakes across a pH gradient (pH 2–4). *Hydrobiologia*, **433**, 3–14.

Woods, M. and Moriarty, P. V. (2001). Strangers in a strange land: the problem of exotic species. *Environmental Values*, **10**, 163–91.

Woodward, F. I. (1993). How many species are required for a functional ecosystem? In *Biodiversity and Ecosystem Function*, ed. Schulze, E.-D. and Mooney, H. A. Berlin: Springer, pp. 271–91.

Wright, J. P., Symstad, A. J., Bullock, J. M., Engelhardt, K., and Bernhardt, E. (2009). Restoring biodiversity and ecosystem function: will an integrated approach improve results? In *Biodiversity, Ecosystem Functioning, & Human Wellbeing: An Ecological and Economic Perspective*, ed. Naeem, S., Bunker, D. E., Hector, A., Loreau, M., and Perrings, C. Oxford: Oxford University Press, pp. 167–77.

Wright, L. (1994). Functions. In *Conceptual Issues in Evolutionary Biology*, ed. Sober, E., 2nd edn, Cambridge, MA: MIT Press, pp. 27–47.

Yachi, S. and Loreau, M. (1999). Biodiversity and ecosystem productivity in a fluctuating environment: the insurance hypothesis. *Proceedings of the National Academy of Sciences*, **96**, 1463–8.

Yaffee, S. L. (1996). *Ecosystem Management in the United States: An Assessment of Current Experience*, Washington, DC: Island Press.

Yellowstone National Park (1995). *Resource Management Plan*, Mammoth Hot Springs, WY: National Park Service.

Yellowstone National Park (1997). *Yellowstone's Northern Range: Complexity and Change in a Wildland Ecosystem*, Mammoth Hot Springs, WY: National Park Service.

Zavaleta, E., Pasari, J., Moore, J., Hernández, D., Suttle, K. B., and Wilmers, C. C. (2009). Ecosystem responses to community disassembly. *Year in Ecology and Conservation Biology 2009*, **1162**, 311–33.

Zenetos, A. (1996). Classification and interpretation of the established Mediterranean biocoenoses based solely on bivalve molluscs. *Journal of the Marine Biological Association of the United Kingdom*, **76**, 403–16.

Zerbe, S. and Wiegleb, G. (eds) (2009). *Renaturierung von Ökosystemen in Mitteleuropa*, Heidelberg: Spektrum Akademischer Verlag.

Index

Printed in the United States
by Baker & Taylor Publisher Services